Himachal Pradesh : Showing Kinnaur District

About the Author

SHIVA CHANDRA BAJPAI (*b.* 1931-) took his M.A. degree in History in 1958 from Lucknow University and Ph.D. in 1969 from the University of Saugar. He was a student and UGC Fellow at the Indian School of International Studies, Jawaharlal Nehru University, New Delhi From 1973 to 1977 he was Visiting Fellow at the Indian Institute of Advanced Studies, Shimla. His publications are, *The Northern Frontier of India* 1970, contributions in some other publications and research papers and articles in leading academic journals, on various aspects of Himalayan and Central Asian regions.

Dr. Bajpai was engaged in research on Himalayan Frontier Regions of India as an ICHR Fellow attached to the School of International Studies, Jawaharlal Nehru University.

KINNAUR IN THE HIMALAYAS

KINNAUR IN THE HIMALAYAS

Mythology to Modernity

S.C. BAJPAI

CONCEPT PUBLISHING COMPANY PVT. LTD.
NEW DELHI-110059

The Indian Council of Historical Research, New Delhi, gave the financial support towards the publication of this work. The Council, however, is not responsible for the views expressed therein.

ISBN : 978-81-7022-180-7

Edition 2021

Published and Printed by

Concept
Since 1974

Concept Publishing Company Pvt. Ltd.
A/15-16, Commercial Block, Mohan Garden,
New Delhi-110059 (INDIA)
T : +91 11 41101460
E : publishing@conceptpub.com, *W* : www.conceptpub.com

Cataloging in Publication Data--Courtesy: D.K. Agencies (P) Ltd. <docinfo@dkagencies.com>

Bajpai, Shiva Chandra, author.
Kinnaur in the Himalayas : mythology to modernity / S.C. Bajpai.
pages cm
Reprint. Originally published: 1981.
Includes bibliographical references and index.
ISBN 9788170221807

1. Kinnaur (India)--History. 2. Kinnaur (India)--Social conditions. 3. Kinnaur (India)--Economic conditions. 4. Kinnaur (India)--Politics and government. I. Title.

LCC DS485.K45B35 2021 | DDC 954.52 23

FOR

MRS. CHANDRA MOHINI BAJPAI

Preface

INDIA'S Himalayan Frontiers, comprising ranges of varying elevations, occupy a unique position in her history. In legend and mythology they were the abode of the gods and hence worshipped; their lonely splendour provided a haven for saints to do their penance; they were often the refuge of defeated or conquered peoples; again they were obstacles to be overcome by marauders in medieval times; while today their solitary peaks offer a challenge for mountaineers. Himalayan ranges are no longer intractable in this age of space shuttles. In the present day tension-torn world the Himalayas have acquired a great importance to us from the viewpoint of security, strategy, integration and knowledge. Hence an over-all view of the Himalayan frontiers is essential, and a detailed study of its regions most desirable.

The Kinnaur region is situated in Himachal Pradesh of India, on the international border with the Tibet region of China. Kinnaras, the inhabitants of the region, have deep roots in Indian mythology, legend, and literature. Mythology tells us that it was the abode of the Pandavas during their year of hiding (*aqyatwas*) while in exile, and the origin of the institution of polyandry is traced back to the legend of Draupadi who had five husbands (the Pandava brothers). Perhaps with the sole exception of the *Rig Veda*, all ancient literature, be it the *Vedas* or *Puranas* or the legal work of Manu (*Manusmriti*) or the literary work of Kalidas (*Raghuvamsha*) or religious works like the *Ramayana* and *Mahabharata* has referred to the Kinnaras in some form or the other. Yet, perhaps no systematic and comprehensive work is available on the people or on the region. This study is just an attempt in this direction.

The virtually inaccessible terrain and inhospitable climate had been conquered by the Kinnaur people inhabiting this region. What kind of a people are they? What of their language, religion, customs and rituals? How do they clothe and adorn themselves; what kind of houses do they live in, and how do they entertain themselves? What is their family structure; what are the moral values that go with their life style? What of their history—past and present; and what has been their economic development through the years?

The very sustenance of the people depends upon their economy. The Kinnaur people have, through the ages, come to acquire several skills as in the fields of agriculture, animal husbandry, horticulture, forestry, cottage industry and trade. Kinnaur polity has developed from primitive times through *goris*, *waziris* and kingship to modern times. What has been the impact of modern systems of communication, educational and medical facilities, the coming of electricity to the region as well as social services? How far-reaching a change have these made on the economy and outlook of the people ? These are some of the relevant topics that have been studied. An attempt has also been made to compare Kinnauri culture with those of some other parts of the country.

The source material on the subject is scattered and scanty. I have tried to locate and utilize almost all published and unpublished material preserved in the National Archives of India, New Delhi; Himachal Pradesh Secretariat Library, Simla, Census Department Library, Simla, Indian Institute of Advanced Study Library, Simla and other libraries located at Simla. Extensive field observations, long conversations with the local inhabitants, and participation in the fairs, festivals and ceremonies of the area have helped me a great deal in understanding the habits, nature and impulses of the Kinnaur people.

I have tried to treat the subject chronologically. That is to say, what it was in the ancient past and how it has reached the present stage of development. I regret that only partial success has been achieved in this direction.

The problem of spelling of proper names is very difficult as they are variously spelled. There are generally two sets of spellings of most places, such as Puh, Pooh; Morang, Moorang and Nachar, Nichar. I have taken only those spellings which are nearer to the pronunciation of the people, such as Puh, Morang and Nachar.

The book was written during my Visiting Fellowship at Indian Institute of Advanced Study, Simla.

I gratefully record my obligation to all those *Non-cooperators, Satyagrahis* and *Security Prisoners,* who fought for the freedom of the country, and have encouraged and inspired me from early childhood in the direction of learning and research. I owe a great debt of gratitude to all the scholars who have written on the Himalayan regions, and upon whose writings I have drawn liberally. I am indebted to Prof. S.C. Dube for the keen and consistent interest he has shown in the completion of this study. I am thankful to several authorities of Himachal Pradesh Government, particularly the late Dr. Y. S. Parmar, former Chief Minister, and Mr. Attar Singh, I.A.S., for providing facilities at Simla and Kinnaur. I am thankful to Shri Ram Saran Negi of Village Telangi and Shri Y. C. Negi of the Census Department for their help in understanding the Kinnaur society and customs. I thank the library staff along with Miss Aleka Mukherjee, at the Indian Institute of Advanced Study and Mian Goverdhan Singh of Himachal Secretariat Library for their constant help to me. Several friends at the Indian Institute of Advanced Study have been helpful to me during my stay at Simla. My thanks are due to all of them. I am thankful to Shri Ashok Kumar Sharma and Shri S.A. Jabbar of the Institute who have procured materials at short notice for the publication. I am grateful to Dr. Biswajeet Sen who has provided some photographs for this work and to Shri D.C. Sharma who has edited the manuscript for the press. Finally I express my most profound feelings of gratitude to Shri Markandey Dubey, Mrs. Gargi and Ravindra Vasishtha, for all the care they took for my comfort, when it was needed most.

As an amateur in inter-disciplinary studies, I am conscious of my shortcomings. Errors and omissions, if any, are mine and the responsibility for ideas, comments and judgements rests solely on me. I crave the indulgence of my readers.

The book is dedicated to my wife who alone suffered the agony during my long stay in Kinnaur, and during the anxious period when it was rocked by a severe earthquake.

New Delhi S.C. BAJPAI

August 1981

Contents

PREFACE vii

LIST OF PLATES xii

1. ECOLOGICAL SETTING 1

2. SOCIAL SETTING 9

3. HISTORICAL PERSPECTIVES 46

4. ECONOMIC RESOURCES 69

5. POLITICAL ORGANIZATION AND ADMINISTRATION 92

6. DEVELOPMENT AND ATTITUDES 111

7. REFLECTIONS AND CONCLUSIONS 130

APPENDICES 146

i. Household Utensils *146*
ii. Ornaments *148*
iii. Kinship Terms *150*
iv. Hindu Gods and Goddesses with their Location *152*
v. Fairs and Festivals with their Location and Duration *154*
vi. Genealogy at Kamru *167*
vii. Trade between Bussahir and China Communicated by Moonshie Currumooddin-1815 *171*
viii. Captain Kennedy's Report on the Protected Hill State-1824 *177*
ix. Kinnaur in Glossary of Tribes and Castes in Punjab and NWFP based on Census of 1881 and 1891 *204*
x. Glossary of Kinnauri Language *218*

BIBLIOGRAPHY 233

INDEX 241

List of Plates

BETWEEN PAGES 16-17

1. Kinnaur lady
2. A typical working lady from Kinnaur
3. Kinnaur belles from the three regions
4. Kinnaur jewellery

BETWEEN PAGES 32-33

5. A typical house in Kinnaur
6. Kalpa Bazar
7. Village headman in Buddhist monastery
8. Kinnaur school teacher with a lady
9. Kinner Kailash
10. Kinner Kailash

1

Ecological Setting

KINNAUR REGION, formerly a part of the native State of Bashahr, was until recently known as Chini tahsil of Mahasu district. On 1 May 1960 it came into being as an independent district comprising Chini tahsil and fourteen villages of Rampur tahsil of the earlier Mahasu district which now forms a part of Simla district. It is so constituted on account of its peculiarities of language and other natural factors. Thus, Kinnaur, the north-eastern frontier district of Himachal Pradesh and a border district of India lies in the Western Himalaya on both the banks of the river Sutlej. It is situated between 31° 05′ 50″ and 32° 05′ 15″ North latitude and between 77° 45′ and 79° 00′ 35″ East longitude. It is about 80 kilometres in length and about 65 kilometres in breadth.

BOUNDARIES

Kinnaur is bounded on the east by the Ngari region of Western Tibet; the district is separated from Tibet by the Zadskar Mountains. The Dhaula Dhar range of mountains forms its southern boundary and it separates Uttar Kashi district of Uttar Pradesh and Rohru tahsil of Simla district from it. Srikhand Dhar separates the district from Kulu and Rampur regions in the West. In the north Kinnaur district is separated from the Spiti region of Lahaul and Spiti district by the rivers Spiti and Pare near the international boundary with Tibet.

PHYSICAL FEATURES

The face of the district presents high hills and low dales, with rapid and rushing streams and streamlets and is marked by precipitous sky-high mountains with their peaks perpetually covered with snow. The district is almost equally divided by the main Sutlej Valley. The River Sutlej enters the district from Tibet in the northeast near the village of Namgia and leaves it at the western end at Chauhra near Wangtu bridge. The district is situated at a height between 1,220 and 3,050 metres above the sea-level. River Sutlej enters the district at an elevation of about 3,050 metres and leaves it at the elevation of 1,220 metres. There are three more or less parallel mountain ranges: the Zadskar mountains, the Great Himalaya and the Dhaula Dhar. The crest of the Zadskar mountains forms the eastern international boundary of Kinnaur with Tibet, the great Himalaya extends from the north west to the south and the crest of the Dhaula Dhar constitutes the southern boundary of Kinnaur at the south eastern corner where the last two ranges merge. The general elevations of the peaks on these mountains vary between 5,180 and 6,770 metres, and hence are covered with snow all the year round. The chain of the snow is broken only at the passes which allow the passage from one drainage system to another. The Zadskar mountains which rise from the River Spiti have the distinction of having two highest peaks, Leo Pargial 6,770 metres and the other 6,608 metres in the district. In the great Himalaya at the Sutlej and Spiti divide rises the third highest peak 6,593 metres and overlooks Mauirang Pass at the head of the Ropa Valley. Kinner Kailash or Raldang rises from the base of Sutlej River at an elevation of 1,900 metres to spectacular 4,573 metres in 'a sheer wall of the ice and snow, overlooking on the opposite bank of the district headquarters, Kalpa'.[1] The beauty of Kinner Kailash from the P.W.D. guest house at Kalpa can be viewed day long with changing colours on account of the movement of the sun rays on the snow. The Dhaula Dhar range has peaks ranging generally from 4,877 to 5,791 metres. Between Zadskar, the great Himalaya and Dhaula Dhar mountain ranges lie several valleys of varying dimensions. The variation of these are from glens and ravines to sizable valleys; the former may constitute Tirung and the latter Spiti and Baspa. The district is formed of great masses of mountain spurs, and the only level ground to some extent, is the Baspa Valley at an elevation of 2,438 to 2,591 metres.

VALLEYS

Among the valleys, Sutlej Valley is the largest embracing both sides of the river. The general direction of this valley is northeast to southwest with its length about 140 kilometres. On either side of the river Sutlej there are several streams and ravines exhausting themselves into the main stream. Extensive vineyards and thriving crops, interspersed with orchards of apricots and apples are found here. The right bank of the river contains better pasturage. Hangrang Valley is the second largest valley, situated on the uppermost part of the district bordering Tibet and Spiti. It is about thirty-two kilometres in length, spread on both sides of the River Spiti. Hango village with its population of 404 is situated at an elevation of about 2,800 metres and Sumrah with its population of 154 is situated at an elevation of about 3,660 metres. The point where Spiti river meets the Sutlej river is extremely rugged and highly dangerous. There are only eight villages such as Sumra, Shyalkhar, Hango, Chuling, Nako, Chango, Malling, and Leo in the whole of this tahsil. The face of the Valley by and large exhibits a picture of sterility and is very different from the evergreen forests of the lower tracts. The fields at an elevation of 3,960 metres at Nako along the River Spiti yields abundant harvests and are fit for vineyards only.

The next is Ropa or Shyaso or Sangnam Valley formed by Ropa stream. This stream passes through an inhabited area of about thirteen kilometres of which about eight kilometres is rich for cultivation. It possesses orchards of apple, apricot and fields of vineyard where water is available in abundance. The rest of the Valley has very little wood except stunted pine and birch. The important villages in this valley are Ropa, Giabong, Sangnam, and Shyaso. Next to Ropa in importance is the small wooded valley of Wangpo or Bhabha, through which a torrential stream runs and merges in the Sutlej. There are about nine villages, namely, Katgaon, Huri Karaba, Yangpa, Bei, Yutrang. Siangeon, Kafnun, and Homte. The Tidong Valley through which the Tirung stream runs is the most rugged with hair-raising pathways along the banks of the stream and has only thirty-two kilometres of cultivable land along the stream. Due to its being rugged in nature with full of rocks, caves, and due also to the lack of means of communication with the outside world, the population of the valley is very low. It has the distinction of having the highest village of the region, Charang at 3,600 metres with a population of 237. It has two more villages, Thangi and Kunu.

The Baspa or Sangla Valley is the most beautiful and romantic in Kinnaur district. The length of the river in the valley is about ninety-five kilometres. On either side of its course are situated most of the important villages of this district. The valley is richly cultivated. It has green pasture lands on both the sides and the meadows nearer the valley are full of flowers and fruit trees. At the head of the river lies the lofty Chung Sakhago Pass. The upper part of the valley, almost half of its total length and up to the village Chitkul (3,475 metres), there are barren ranges covered with snow all the year round. The lowest point in the valley is 1,830 metres, where it meets the Sutlej. From Chitkul to Sutlej is the inhabited part of the valley.

Among the valleys there are a few more such as Gyanthing or Nesang, Pejur or Leppa, Kashang, Mulgoon and Yula. These valleys have rich grazing lands, a boon for sheep and goats.

RIVERS AND OTHER WATER RESOURCES

Sutlej the principal river of the district has its origin in the Himalayas and has plentiful and perennial sources of water. It enters Kinnaur district from Tibetan territory by a pass and passes through the district, almost dividing it into two parts. Within the district the length of the Sutlej is about 130 kilometres. Another important river of the district is the Spiti or Lee. It has its source in the far north on the eastern slopes of the mountain ranges which run between Lahaul and Spiti. Before meeting the Sutlej at a place Khab at an altitude of 2,589 metres, Spiti is joined by several feeders which meet the river both at right and left banks. On the left bank it is met by a torrent Chaladopko and on the right side of it are the Julang, the Lipak, and the Tirasang. The Baspa River, another feeder of the Sutlej rises from Dhaula Dhar mountain of the Himalaya.

passes through the valley bearing name and meets the Sultej at a place called Karcham, after a distance of about seventy-two kilometres. Several feeder streams feed the Baspa river before it meets the Sutlej. On the left Zapkia, Thatang, and Bering meet it, and on the right is the Suthi. Besides these three main rivers there are several other small streams which fall into the main Sutlej river, such as Mulgoon, Yula, Wanger, and Tirung. Several *khads* also increase the volume of water in the river Sutlej from both banks. On the left bank the important khads which meet the river are Baturing Khad, Brua Khad, Chitkue Khad, Harba Khad, Rasorag Khad, Shong Khad, Shoni Khad, and Singon Khad, whereas on the right side are Gorarung Khad, Rona Khad, and Sushang Khad.

River water is the main source of water and besides this there are two small tanks located at village Nako in Hanarang Valley and the other a *cutcha* tank or *sorang*, as it is called locally, is situated above the villages of Janu and Ramani in tahsil Nachar. Several springs and spring heads are scattered all over the district. Another source of water is snow. As has been noticed earlier, the area of the district above 5,200 metres remains permanently under snow all the year round and the area above 4,250 metres is covered by snow for more than six months in a year. The remaining area between 4,250 and 2,000 metres is covered by snow from anything between three and nine months of the year. Winter snowfall occurs all along the district from Chaurha to Smuarah.

During the summer months this snow melts and the water yielded by it is mostly absorbed by the surface of the hills. This absorbed water gradually percolates and ultimately forces its way out of the ground and forms a spring locally called *dhara*. Generally this source of water lasts the whole year or till the next winter comes and there is more snow. Some of the springs die out altogether before the onset of the snowy season. Generally speaking, however, the water is available out of the melting of winter snow all the year round.

It is popularly believed by people that there is some sort of warm current flowing along the course of the river Sutlej which keeps the temperature warm, suitable for paddy cultivation. This popular belief is supported by the occurrence of several hot springs in the district. One such hot spring is located at the village Natpa at a distance of five kilometres from Nachar in the north and another is available at Tapri the present major bus centre in Kinnaur. There is yet another at Joktiang hamlet, about five kilometres from Nachar in the east. The waters of these hot springs are very useful to the people for bathing and washing of clothes.

CLIMATE

Kinnaur, due to geographical situation has a long winter from October to May, the snowy season, and a short summer from June to September. From April to May is spring and from September to October is autumn. It has two distinct climatic zones—the wet and the arid. Only the area south of the great Himalaya receives monsoon rains. This area roughly covers Baspa Valley and the lower reaches of the Sutlej Valley. In the upper reaches of the district the monsoon showers progressively decrease and one can notice the beginning of the complete arid zone from the place called Spilo at a distance of 255 kilometres from Simla, situated at an elevation of about 2,276 metres above the sea level, Kalpa the district headquarters may be considered as a place where the great Himalaya obstructs the march of the monsoon. The vegetation above Kalpa rapidly becomes sparser ultimately to remain only along the river. Along the valley of the Sutlej as far east as Wangtu the rainfall does not greatly vary from Simla, but beyond Wangtu the difference in considerable. The rainfall becomes less and less as Shipki is approached so that the climate of the upper Kinnaur is semi-arid. West of Wangtu, the Sutlej Valley has an annual rainfall of 175 centimetres. At Kilba, 16 kilometres east of Wangtu, this drops to 107 centimetres, and at Puh to 40 centimetres. In the lower portion of Kinnaur, June is the warmest month and in the upper regions July and August are the warmest months. From 2,500 to 3,000 metres in a favourable situation such as at Chango, Leo and Morang the temperature of July and August ranges from 20° to 23°C, the October temperature is about 10°C. At 3,500 metres the summer temperature appears to be from 13°C to 3°C. May and October have very low relative humidity and are generally the drought months. Clouding is more heavy, persistent and prolonged in the long winter season. There are strong winds in the winter months. The area is free from storms of great intensity such as cyclones. Thunderstorms occur very rarely here but dust storms never occur in the region.[2]

AREA AND POPULATION

The total area of this great and peculiar land mass is 6,553 square kilometres. Its total population is 59,154, with an average of 9 persons per square kilometre. The entire population is classified as Scheduled Castes or Tribes. There are no urban centres in the district and the total population lives in seventy-seven revenue villages, scattered in six divisions, such as Hangrang, Puh, Morang, Kalpa, Nachar and Sangla. Nachar division is the most populous with a total of 14,205 persons and Hangrang the least with only 3,125 heads. The population of the district is composed of four distinct classes, of Kanets (Rajputs), Domangs (blacksmiths), Ores (carpenters) and Chamangs (kolis).[3] Compared with the plains population of the villages these classes are, Rajputs, Lohars (blacksmiths), Barahi (carpenters), Kori (weavers), and Chamars (cobblers).

OCCUPATION

Majority of the population comprises cultivators. Next to this group, in numerical order, come those engaged in household industry, in construction, and other services. Some people are engaged as agricultural labour, in mining, quarrying, livestock, forestry, fishing and hunting. There are some people who are engaged in plantations, orchards and such other activities. There are several who are busy in manufacturing industries other than household industry and trade and commerce. A few work as transport workers and labourers. Nowadays most people work as labourers in the Public Works Department on daily wages.

COMMUNICATIONS

In the rugged mountainous district, the means of communication constitute the greatest problem. In the past, communication was limited to high passes and the river valleys. The important passes in the district are Shipki, Lamkhaga, Borasu,

Nalgan, Buran, Shathal, Shakarog, Khango Tari Khango and Manirang. The height of these passes is between 5,500 metres and 4,600 metres. There was only one important road that led from Simla and entered Kinnaur beyond Sarhan. In Kinnaur the road passed through the villages of Taranda, Nachar, Wangtu, Chagaon, Pangi, Kanam, Sangnam, Hango, Puh, and Shipki and thereafter entered into Tibet. This was the Hindustan-Tibet road, and was the main line of communication between Tibet and Kinnaur. Trade was conducted through it.[4] After independence, particularly after 1962, road construction has spread a network of various types of roads in the district. Besides National Highway No. 22, there are link roads between this highway with Kalpa, and Nachar. Other jeepable roads are the Upper Hindustan-Tibet Road, the Karchham-Sangla-Chitkul Road, the Taiti Valley Road, the Ropa Valley Road, the Namgia Link Road and the Kilba-Shaung Khad Road via Karchham. There are about fourteen bridle paths scattered all over the district, which serve the purpose of communication today.

REFERENCES

1. M.D. Mamgain, ed., *Himachal Pradesh District Gazetteers* : *Kinnaur* (Ambala, 1971), p. 5.
2. For ecological details, reference can be made to S.C. Bose *Geography of the Himalaya* (NBT, New Delhi, 1972); S.P. Chatterjee, *Physical Features of the Himalaya*, Papers and reports of the Seminar on the Himalaya, December 1965, I.S.I and Delhi School of Economics (New Delhi), pp. 17-19; R.L. Singh, ed., *India—A Regional Geography* (Varanasi, 1971), p. 440, *Census* 1961 District Handbook No. 6 *Kinnaur*, pp. 3-8; M.D. Mamgain, ed., *Himachal Pradesh Distict Gazetteers*: *Kinnaur* (Ambala, 1971), pp. 1-47.
3. *Census of India 1971:* Series India Paper I of 1972. All the figures are drawn from the Census of 1971. Only total population figures are from *Census of India, 1981*, Series 7 Himachal Praders, Paper-1 of 1981.
4. Communications in the early nineteenth century are given by Captain Alexander Gerard in his book, *An Account of Koonawar in the Himalayas etc. etc.* (1841), pp. 36-7.

2

Social Setting

THE INHOSPITABLE terrain of Kinnaur district is inhabited by the hardy people of the Kinner tribe. Some people believe that they are the same as the Khash of Western Himalaya. It appears that the Kinners are distinct from the people of other areas of Western Himalaya. Their location is limited to the district of Kinnaur particularly on either side of the river. Here they are known as Kanets. Besides this majority tribe there are several other people belonging to the Scheduled Castes.

The people along the Tibetan border possess the Mongoloid and Tatar features, the rest of the tribals are of a fair complexion and good looking with Persian features. They are physically well built and muscular and are between five feet and five feet ten inches in height. The people, 'are frank, active, generous, hospitable and highly honourable in their dealings.'[1] They lead a hard life and their habits are simple, and wants limited. They are enterprising and travel far and wide in pursuit of trade and commerce and have acquired a reputation for honesty in their dealings. A traveller of the early nineteenth century has observed:

> . . . that they are the frankest and most honest race of people I have seen in India. They possess neither craft nor ingratitude, and they may be trusted to the world's end. Thieves and robbers, the indigenous inhabitants of mountain regions are unknown and the same character belongs to the interior tracts of Busahir, or to that portion of the interior called Koonawar laying within the snowy mountains.[2]

LANGUAGE

The Kinnauri dialect, called Homskad, is the mother tongue of nearly seventy-five per cent of the population of Kinnaur. There are as many as nine different dialects used by various sections of the population. One of them Sangnaur is spoken in a solitary village Sangnaur of tahsil Puh. The villages on the Tibetan border speak Tibetan dialects spoken in western Tibet. The extent of spoken Tibetan is limited to the village of Nesang Kune and Charang adjoining Tibet. Jangiam dialect is spoken in Jangi, Lippa, and Asrang villages of Morang tahsil. The Shumccho dialect is spoken in the villages of Kanam Labrang, Spilo, Shyaso, Tailing and Rushkalang, of Puh tahsil. A Kinnauri-Jangram mixture is the language used in Rakcham and Chitkul villages of Sangla tahsil. The Scheduled Castes or Harijans speak a language which is closer to that of certain parts of the adjoining districts of Kinnaur. Besides these dialects the educated people of Kinnaur can speak Hindi also both men and women, specially in Sangla and Kalpa valleys, can speak English in addition to their mother tongue and Hindi.[3]

CASTES

There are two principal castes in Kinnaur district. They are Kanets or Rajputs and Scheduled Castes. Among the Kanets there are three grades. In the first grade Kanets there are as many as fifty sub-castes, in the second grade there are seventeen sub-castes, and in the third grade who work as potters have three sub-castes.[4] Waza Kanets belong to the third category and are considered inferior among Kanets. Several sub-castes of Kanets are akin to sub-castes among Brahmins and Rajputs or Hindus as a whole in the hills and plains of India.

The Scheduled Castes include Chamang and Domang. Chamangs make and mend shoes and weave cloth. The Domangs are primarily blacksmiths. There is a third caste called Ores. The main profession of Ores is carpentry. In social status Ores are equal to the Domangs. Among Scheduled Castes, blacksmiths and carpenters, i.e. Domangs and Ores consider themselves superior to Kolis or Chamangs. Domangs and Ores do not eat with Chamangs and do not enter into matrimonial ties with them. Similar customs are prevalent among the Scheduled Castes people of the plains of northern India, particularly U.P., Bihar, and Madhya Pradesh.

Among Kanets or Rajputs, the general restriction is that water or cooked food touched by the lower castes is not used by them. If a Kanet or Rajput ate such food inadvertently, he, during the princely regime would apply to the Raja who bade him make expiation and pay some *nazrana* or forfeit. This custom was called *sajeran* or *sucheran*. Among Kanets intermarriage within the sept is strictly speaking, not allowed but inter-sept marriages do take place, i. e. for instance the high Kanets and Wazas marry among their own castes only. The girls of Wazas are not taken in marriage by the Kanets nor do they give their daughters to Wazas. If any one out of higher Kanets marries a girl from Waza caste he is considered inferior and ex-communicated. Similarly, if a girl out of Kanets contracts a marriage with a Waza boy she is considered inferior and ex-communicated. However, the redeeming feature in this kind of restriction, is that if there are such marriages in three successive generations in the high caste families, the offsprings of that family attain the status of the family of high caste in the third generation without reference to society. Closely akin to the system of *bigha*, *biswas* in Kanyakubja Brahmins of north India, a gradation system of high and low is denoted by a terms locally pronounced as *orung* among Kanets. People belonging, to the same *orung*, though they may come from different sects can marry among themselves. They can eat together in the same utensils, which is not permissible in case they belong to different *orungs*. When a man of lesser Orung or no Orung eats in the house of one who has higher orung, the former has to wash his own utensils etc. It is difficult to trace the origin of orungs. Perhaps the plains people who came to Kinnaur and settled from time to time have brought it from there and have asserted their superiority over local people. The practice of *gotra* and *viswas* (a degree of purity) is prevalent in the plains, particularly in U.P., Bihar, Madhya Pradesh and Rajasthan. Hence a close similarity can be established with the people of Kinnaur and the people of northern India.

These traditional restrictions are based on the caste system and have been playing their role. But after independence the hold of caste rigidity and consequent restriction of marriage alliances and eating is on the wane though not completely extinct as yet. Perhaps it will take a long time to eradicate caste distinctions effectively.

Despite legislation preventing untouchability the social customs have retained it. The Chamangs and Domangs cannot mix freely with the people of other castes. There are differences among the lower classes of Kanets and Rajputs. The Chamangs whose main occupation was connected with the removal of dead animals have of late, withdrawn themselves from this occupation though they still plough the fields and prepare shoes from leather got from the plains. The caste problem is one of the major problems of present day India. In recent times the caste rivalry has led to several armed conflicts in U.P. and Bihar particularly.

HOUSING PATTERN

As the climatic condition between upper and lower Kinnaur varies so also does the housing pattern. Lower Kinnaur has an abundance of forests and rain while Upper Kinnaur is deficient in both. But one thing common to both is that all the houses are built facing the sun.

Generally speaking most houses in Kinnaur area are of two storeys. But the houses of nobles or landlords comprise more than two storeys. Except in the new constructions, each storey has only one living room; usually the ground floor is used for cattle sheds, except where there are workshops as in the case of ironsmiths. The first storey is used for living purposes. It has extended wooden balconies. The material used for the construction of houses is different in lower and upper Kinnaur. Generally stone and wood are used. In the lower Kinnaur, where wood is available in abundance, the wooden frame is made first and the gaps are filled with stones. Mortar was not used in earlier times. In the upper Kinnaur where wood is not readily available the houses are made of stone. But after the recent devastating earthquake (1976) the wood is carried from lower Kinnaur to the upper region. Now one can see wooden houses in villages like Chango and Sumruh. In

Kinnaur there are low-roofed houses where the height of the wall does not ordinarily exceed two metres. Great significance is attached to the main beam of the roof. It is laid almost parallel to the boundary of the main door. It is considered inauspicious to set it otherwise. There are only small holes for letting out smoke. Windows and ventilators were not provided in earlier constructions of the region due to the extreme cold winters. As in other parts of the country, doors open inwards and in some cases they can be folded and are fitted with chains (*sankar*) which are fixed to the balcony above. In the plains these chains are fixed to the outer frame (*chaukhat*). Since each storey comprises only one room, there are no separate bathrooms or kitchens. The hearth called *mieling* is set in the centre of the room and a hole called *dusrang* is provided in the room for the dual purpose of letting the smoke out and admitting the light. A handy stone or a wooden plank is always kept ready to close this hole during rain or snow. The inside of the houses is plastered with a mixtuie of mud and cow-dung, and is white washed by a shiny kind of mica. The roofs of the houses are mostly flat and are made of wooden planks covered with *bhojpatra* (tree bark) and overlaid with earth. The floors in lower Kinnaur are wooden while in upper Kinnaur they are plastered with mud and cow-dung, as wood is warm in winter. An *urch* (granary) for food grains is constructed at a distance from the main house, to avoid destruction by fire or house collapse.[5]

At the time of laying of the foundation stone and fixing of the main door, worship with *pole*, *porsad* (offerings) and some wine is offered near the foundation and the door. On the completion of the new house *gorasang* (*grahapravesh* or housewarming function) is held on an auspicious day and hour. After worship and chanting of some sacred verses, the lama formally escorts the owner inside the house. Similar customs are observed in the rural areas of the plains, with the only difference that in place of the lama, a Brahmin performs the ceremonies. In Kinnaur, as elsewhere in the plains, the mason who constructs the house is invited to the house warming (*grahapravesh*) or *Gorsang* ceremony for a feast. In earlier times the mason and his family were given gifts and were treated very courteously. On the morning he was generally given a *nang* (*thali* or plate) with ghee filled up to the brim along with five or seven rupees. The mason and his wife were also given clothes.

His wife, in addition, was given a pair of *dhaglo* (bangles). A great care was taken to please the mason on this occasion, as it was feared that he might otherwise curse the family should he remain unsatisfied. This was however, an old custom. Today carpenters and masons are employed on daily wages.

FURNITURE, UTENSILS AND DECORATIONS

The use of cots or beds in Kinnaur is very rare. Except for a few rich, people by and large sleep on the floor on the mats and skins of goat or sheep which keep them warm. These mats are locally called *khera* or *kharacha* and *pakpa*. Among the household goods a *kote* or *kotich* (a wooden chest) was the most important. It was used for keeping grain or apricots. Next in importance was the basket slung on the back. It is even now used for carrying goods from one place to another. A wooden loom was a must for every house. Among other things a stone mortar and pestle was kept for extracting oil from the kernel of the apricots and walnuts, a hand mill for grinding grain, a *hookah* for smoking; while skins of various types and sizes for stocking flour and grain, for carrying water, butter and milk, were kept in almost every household in Kinnaur. Even now skins are used for sitting, but now carpets are also used. Clothes are kept in wooden boxes and trunks. *Khanda* which is a deep square box is used by many households for keeping grain. It can hold as much as fifty maunds of grain at a time.

Grinding is often done in water mills called *gotang*. Sometimes hand mills are also used. They are known as *hasgotang*. A weaver's apparatus which is very simple is kept by each Harijan household. With the change of time, some well-to-do people in Kinnaur have acquired some items of furniture. These include cots, tables, benches and chairs. The things of the past are mostly replaced by modern ones. Due to the availability of electric power, most of the grinding is now done by power machines.

Brass and bronze utensils are in common use in the district. Aluminium utensils are in beginning to find favour with the people especially during travel. Earthen pots are also used besides some wooden articles. Vessels made of metal are procured from the plains, while wooden or earthenware is made locally. Some well-to-do families keep silver utensils also. Now crockery and dinner sets are also available with the people of means.

Walls are decorated with pictures of men, women, trees, moon, stars, sheep, goats, cows, and flowers drawn crudely with a white chalk-like material. The walls and roofs of temples and Buddhist monasteries are painted in bright colours. Mostly religious motifs are painted. Beautiful wood work can also be seen at these religious places. With the opening up of the area and the advent of modern civilization, some houses are now painted with modern paints radio, transistors, watches, clocks and other luxury items are also available in the area. P.W.D. guest houses are fitted with heaters and geysers.

DRESS AND ORNAMENTS

Because of the cold climate of the area, people wear woollen clothes throughout the year. Both men and women have a common head dress, a woollen cap locally called *thepang*. Outside Kinnaur this cap is popularly known as Kinnauri or Bushahri cap. On the occasion of marriages, fairs and festivals, Kinnauras tuck flowers in their caps which really add to their charming beauty. The men wear woollen shirts (*chamu-kurti*), long coat (*chhubha*), woollen pyjamas (*chamu-suttan*) mostly of grey colour. During marriages and festivals some people put on *tapru-suttan* embroidered in various designs and colours below the knees. Ladies wear woollen sari (*dhori*) full sleeved blouse (*choli*), *chhanli* or shawl. It is wrapped round the shoulders and its two ends are fastened together near the breast by means of a silver hook called *digra*. Men also wrap the *chhanli* around them during inclement weather. Women often wrap around their waist a *gachhang* a scarlet coloured woollen or cotton cloth of about five to eight metres in length and about a metre in width. Dhoris are of three types. The best and the most costly is called *tapra-se-dhori*. The next best is *arang-se-dhori* and the most common one is *phatta-orang*. All the three types are generally dark or grey in colour and have beautiful patterns in various attractive colours. In place of shoes, Kinnauries wear a local footwear commonly known as *Balzanu-Pona* made of wool and goat hair. Children clothes are made of fine wool and are soft. Now mill made cloth has come to be used increasingly and most females are now clad in *salwar* and *kamiz*, particularly in the areas beyond Puh. Similarly men are in trousers, shirt and coat of modern cut. In some decades the whole pattern of the local dress may undergo a radical change and may even be lost for ever.

Ornaments are the most valuable possession of a Kinnauri woman. They are generally made of silver and include *mool*, *trimani kantaie*, *kandoch chalim shanglavang*, *digra*, *tamach*, *laung*, *dhaglo* etc. Except for *murki* for the ear and rings for the finger, men generally do not use ornaments. Now there is a growing tendency among the women to use comparatively lighter ornaments and thus silver ornaments are getting out of fashion.

FOOD AND DRINKS

Kinnauries have three meals in a day—in the morning, in the noon and at night. Their staple food is wheat, *ogla*, *phafra* and barley. Besides these *kangni*, *(kawing) cheema*, *maize (chhallia)*, *koda (kodru) chollaie dhankher* and *bathu*, are also used. The principal pulses consumed are peas, black peas, beans, *mash* and *masur*. The vegetables consumed are cabbage, turnip, peas, beans, pumpkin, potato and tomato. A mushroom locally called *rangmuch* is taken when in season.

Kinnauris relish meat. Barely two per cent of the population is vegetarian. Goat and ram meat is relished. But chicken is not eaten.

Drinking of alcohol popularly known as (*ghanti*) is common in the district. It is necessary even at the time of offerings to gods (*hawan*). All the communities consume alcohol (*ghanti*) without any distinction. They prepare it for their own use and take a cupful or two along with food. Now a few liquor shops have come up at Kalpa, Nachar and Puh. Smoking is widely common among the people of Kinnaur. A little tobacco is grown here. The rest of it is imported from outside. Even

1. Kinnaur lady (middle)

2. A typical working lady from Kinnaur

(*Facing page 16*)

3. K nnaura Belles from the three regions

4. Kinnaur Jewellery

(*Facing page 17*)

teenagers smoke. The *hukka thoti (chilam)* is preferred but now many are smoking *bidis* and cigarettes.

RECREATION

The extreme climate, mountain-locked isolation and poor economy had prevented the entry of modern means of amusement to this area. As such, people have devised their own methods of amusement. They have strong sense of humour. Besides their participation in fairs and festivals, singing and dancing together is the main mode of entertainment. During the long winter months older people narrate traditional stories and legends of adventures in the inaccessible hills and forests in these areas. Stories of gods and evil spirits are common. The women have very sweet voices and hence have earned the title of *kinner kanthis*. While entertainment goes on people mechanically continue spinning wool for their clothes. *Toshiming* (to sit together) is the most peculiar amusement of this area. This is celebrated all through Kinnaur at different times, and its duration varies from about a week to a month and half in certain areas. *Toshiming* is generally organized by unmarried girls although, sometimes a few married ones also join. To begin with, a party of five to twelve members is formed which is known as *Toshing-Konich*. Among them they choose a leader called *Gomey* and then they decide upon the dates of *Toshing*. A fairly big house in the village is selected for *Toshing*. No rent is paid for the accommodation, though either free labour or free food along with *Toshing-Konich* is provided to the owner of the house. To begin with, these girls start collecting *shadung* (beer) and *phasar* popularly known as *ghanti*. Drinks thus collected are kept at a safe place. Nearer the day of *Toshing* food stuffs are also collected and kept in the house. On the appointed day girls collect together and first clean the house and prepare food. After this they go out in pairs to invite and bring along with them their boy partners, called *Toshing Konich*. They together partake of the food and drink and start singing and dancing. This performance goes on regularly for days together. Musical instruments like *dhol*, *bugle*, *damantu*, *ban* and *shonal* are used but they are played by boys alone. During *Toshing* except most urgent work nothing else is done by the participants.

Sometimes other villagers are also invited to watch the dances. They however, are not allowed to participate, if at times they are allowed to participate in the dance, they are never allowed to share the food etc. Sometimes if a cash award is received by them from the spectators, it goes to the girls only.

> The male partners have to bear the entire expenditure of two meals by providing a ram or a goat and rice sufficient enough for the party. The boys meet this expenditure by mutual contribution. Till the last night of *toshing* dances and songs continue to be performed as usual. The next morning meals are served to the boys by girls. It is now time to disperse. At the time of departure every male partner is presented with a bunch of *chambaka* flowers by girls. In turn the girls are given cash ranging up to sixty rupees, by the male partners. Customarily the girls accompany up to the male partner's house and depart thereafter. The departure after such a gay, joyful and long party is naturally sad and they bid good-bye with heavy hearts.[6]

RELIGION, SUPERSTITION, FAITH AND BELIEF

Mythology tells us that the Kinner people were the worshippers of gods like Kubera, Indra, and Krishna. Perhaps the original religion of Kinnaur centred entirely around the principal village deity, which was a 'divinity, a doctor, a magistrate, a judge, the chief-executive, an astrologer, the village hero, the cynosure of all eyes etc. etc. rolled into one'.[7] In course of time Hinduism of the standard type made an impression upon the people of Kinnaur. This religion had introduced Rama, Krishna, the Ramayana, the Mahabharata, the Gita, the Sukh Sagar, the Prem Sagar, some of the Puranas and some of the relevant systems of worship. Later monastic Buddhism came to be established in the regions bordering Tibet. As a result of this mixing the original form of animistic faith and worship is practically lost. In the mid nineteenth century the religious situation described by Edward Thornton was:

> . . . The religion of Koonawur is Brahminism in the south; in the north, Lamaic Buddhism; in the middle, a mixture of the two systems. There prevails a regularly graduated

> transition from one to the other. Thus, Brahmins are not met with beyond Saharun, near the southern boundary, where they officiate at the shrine of the sanguinary female divinity Bhima Kali, to whom, at no remote period, they offered human sacrifices. At Kanum, about half way between the northern and southern frontiers, the sacred books are in Tibet, and lamas are there first, met with, but Kine are venerated, and some attention paid to the distinction of castes; thus partially amalgamating the two creeds. At Hungrang, on the northern frontier, the religion is pure lamaic Buddhism. . .[8]

The above description was largely true till the early part of the present century. The recruits to the monasteries were boys and girls of Kanets. Even now the girls who do not marry, devote their time studying Tibetan scriptures. They are called *zomos* or *jamos*. They live in nunneries. Almost every village particularly in the north has *zomos*. Likewise boys who learn the Tibeten scriptures, or Buddhist doctrine are called lamas. Like the *zomos* they live in monasteries and are looked upon as holy men. These men are the priests of all the Kanets. There are several monasteries of these lamas in Kanam, Sangnam, Chango, Hango and other villages. Lamas are either *gyolang* (celibate) like the *brahmachari*, or *durpu* who marry but never shave the head. In the case of *zomos*, there is no restriction on their marriage when they get a suitable husband at any stage of life.

At present the situation is that village gods and their retinue still go very strong and, side by side, there exists lamaistic Buddhism and a local version of Hinduism. The Buddhist temple of the Tibetan style and the traditional abode of the village deity stand peacefully side by side or in friendly proximity in many villages.

All the ceremonies, religious or otherwise, are performed by lamas both in Hindu and Buddhist families. The content of the functions are all brahminical in tradition. The absence of Brahmins was noticed by James Baillie Fraser, a traveller of the first quarter of the nineteenth century. His remarks were:

> No Brahmins have ever settled in this district, nor will they go there; perhaps the poverty of the country, and the

> privations necessarily to be suffered during a residence there have deterred these holy men, who usually seem to prefer those places which afford them all the comforts of life.[9]

Why then are the practices brahminical? These people believe in the divine origin of kingship. The Raja of Bashahr was originally from the plains. His religious functionaries were Brahmins, but his entire staff was from the Kinnaur region. These honest people were functionaries at all levels. In course of time they began to imitate the brahminical tradition which spread throughout the region over a period. In the beginning only people of the capital used to perform the ceremonies as were practised at the Raja's palace. In 1882 Sir Denzil Ibbeston had noticed this fact as stated in his glossary of the tribes and castes of this region.

It is strange that no Christians are found in the region. Despite early Christian missions at Puh and other parts of the region, Christianity has not been able to spread very much. The missionary influence is seen in the habits of cleanliness and the use of a refined form of knitting was acquired by the people from the missionaries. Perhaps the ruler of the region who was a Hindu had never encouraged such conversion. Perhaps he even secretly warned the people against conversion. Job opportunities, state patronage and recognition were all in the hands of the rulers of the area, hence the Raja's word would command obedience.

A hard life with difficult terrain and extremes of climate coupled with lack of communication had made people more god fearing and religious. They began to believe in many kinds of superstitions and faiths. These even now continue to affect most of the people of the area. High hills, caves, mountains, inaccessible places like lofty passes, and weird and dangerous spots are inhabited and governed in their beliefs by spirits and beings of supernatural order. The tops of the high hills are supposed to be the abode of Kali and other goddesses. It is a common sight in Kinnaur to see the horns of domestic animals hung at the main door of the house. They are also hung in the ante chambers of the temples. It is believed by the people that these horns have some mysterious virtues. A great virtue is attached to the journey around Kinner

Kailash. It is believed that if a person makes a circumambulation his wishes will be fulfilled. There was a system of secret oath of allegiance which was taken in the name of the supernatural. Originally, probably it was invented by the Raja for the faith and obedience of other people or particularly his employees. This was known locally by the word *darohi*. The oath is occasionally still employed for private purposes. People believe in evil spirits of many types such as ghosts *raskshsas*, *khunkch* etc. Banshir is a ghost believed to live in a forest on the Kail tree. It can assume many shapes, that of a giant, a jackal or monkey and is believed to be a very cruel spirit but not more than Chan who is supposed to inflict sickness on whoever it meets during the odd hours of the night. A man who commits a heinous sin during his life turns into a ghost known as *raskshsas* and it is supposed that he will never attain salvation. *Khunkch* is believed to be a house spirit who can be passed on to another house by selling one's animal or marrying of one's daughter. A *puja* (worship) performed to ward off evil spirits. Similar beliefs are also found in the rural areas of eastern Uttar Pradesh and Bihar perhaps in Madhya Pradesh also. It is believed that he who dies an unnatural death tends to be a ghost or evil spirit. In Kinnaur the full moon period i.e. *Shukla Paksha* is an auspicious occasion for agricultural operations. They prefer Tuesday and Saturday for ploughing and sowing. Before ploughing the earth goddess is worshipped by offering a *chilta* or some *sattu*. A stone is also broken on a crossroad to dispel the evil influences.

Dreams have always exerted a great influence on man's imagination all over the country. In this region, too, the people attach great importance to dreams and they, too, interpret them in their own way. For instance, eating food or killing a tiger in one's dream mean's success in work. During summer if one sees water, a river or a *nulla* in a dream, it is believed that the next day it would rain or be cloudy. In winter if sheep and goats ascend a hill, there would be no rain but when they appear to descend, a snow-fall can be expected the next day. Slaughtering of rams and goats, musicians playing musical instruments, a crow on the verandah of a house or falling of teeth in a dream indicate death. Besides these, putting on new clothes and going out on a horse also points to some death somewhere. Eating green vegetables and fruits is an indication of happiness. Weeping in a dream is a lucky omen. Falling from a wall or tree, laughing or ploughing in a dream are unlucky omens. If a person is seen wearing ornaments or purchasing silver and gold ornaments it is indicative of his death. Eating meat also constitutes a bad dream. If one sees a dog following one should be aware of ghosts and evil spirits. It is believed that if a person has seen a dreadful dream the next morning he should go to temple or a tree and speak it aloud there by which its effects will vanish.

People also believe in certain other superstitions, such as, while going out for some work if one comes across an empty vessel, one is not likely to succeed in the mission. But if a full drum or any vessel is seen then success is indicated. This is true also in the plains, where ladies used to leave a pot full of water just at the door while some one is to proceed for any important work. Similarly, if a person sees a snake catching a *kirla* (chameleon) it is a very bad omen. But if he kills the snake and rescues the *kirla* he has been saved from some bad luck. While going out for some work if an ass is heard to bray it will bring good omen. Unlike as in the plains sneezing is not given any importance here. This may probably be due to the fact that in a cold climate people may be prone to frequent colds. Another bad omen is that if a horse being taken out for riding shivers, it is not considered good for the rider and the journey should be avoided. Similarly, if one has to go out on a long journey and one discovers that the food he is taking with him lacks salt, then the journey should be avoided. One of the most important superstitions is that if in the sky a fiery line called *melup* is seen a fire accident is sure to occur somewhere in the area where it is observed. In the plains the phenomenon is believed as a soul's travel.. In fact scientifically it is nothing but a meteor.

Superstitions, faiths and beliefs are prevalent all over India, perhaps all over the world. Their form, content, and style may differ from region to region and country to country, but the blind faith in them continues despite scientific advancement.

SOCIAL STATUS OF WOMEN

A woman in Kinnaur enjoys a unique position. She is not only the keeper of her hearth but contributes equally, if not more, to outside work. She is an equal partner in dances and takes a leading part in folk songs. In practice women work very hard and from constant exposure to the strain of physical work and weather, soon look much older in appearance than the men. They do all the laborious work, such as carrying loads, weeding and harvesting the crops, carrying water, cooking food etc. In fact, in agricultural operations they do all the work except ploughing. People in the upper regions are priest ridden to a great extent by their lamas.

FAMILY, INHERITANCE AND MORALS

The joint-family system prevailed in general and is still largely in vogue among the people. Under one roof not only father, sons, brothers but also uncles, nephews of the same descent live and sometimes even own property in common. But due to the advent of modern civilization in the area, the system is under serious strain. A number of separation cases year after year indicate that a tendency is developing among the people to live separately in nuclear families and own property.

Inheritance of property is patrilineal. In the case of a polyandrous family when one of the husbands dies his share is inherited by the surviving co-husbands. But when all the husbands die, the sons begotten by them inherit the patrimony in equal shares. In case a person has both legitimate as well as illegitimate children, it is only the legitimate that will inherit the entire property of the deceased fathers. Illegitimate children in the local language are called *poltu*. They marry only in their own class. They live at the mercy of the household as labourers and field workers. Sometimes they are given a field or two not as a right but out of mercy. Such cases are very rare now. In the case of a division of property among the polyandric group, the rule of *jethang* (primogeniture) is followed. In this rule the eldest has the first right to choose then the youngest (*kanchang*). In practice a good field is given to the eldest brother and the ancestral house is given to the youngest. It is believed that the eldest can work hard on the field and get the yield desired and settle himself in life properly. In the case of the youngest, it is felt that as a beginner he should have at least shelter to begin with. The rest of the property is then divided in equal shares. Girls have no legal right in the property even now, as the Hindu Succession Act of 1956 is not applicable to Kinnaur. The right of the transfer of property does not exist in this tribal area.

The morals of the people have been a controversial subject. Until recently monogamy was the exception rather than the rule; polyandry was more prevalent in most parts of the district. In this system several brothers shared one wife. In case a joint wife proved to be barren her sister was brought in as a second wife. Sometimes the youngest brother would prefer to bring another wife for himself on the plea that the common wife was very old. A love affair usually results in separation of one of the husbands from a polyandrous family. In such cases the joint property must be partitioned, unless the new wife consents to be shared by all brothers. In case of refusal, she and her husband have to live separately, but the husband retains his right to share the original joint wife. In practice the original wife refuses to have anything to do with him. With the changes in the region monogamy is no longer an exception. Polyandry is rapidly losing ground to monogamy.

Polyandry is the most debated subject in the area now and opinion is not unanimous as to the usefulness or otherwise of the system. Older people defend it on the grounds that it has kept the family a closely knit entity. It has prevented over-population and the sub-division or fragmentation of small agricultural holdings. It has enabled a family to derive full benefit from several sources for a proper living by way of the pooling of labour. The younger generation disapproves of the system altogether. May be a strong sense of individual possession is the basis for this disapproval.

The legend of Draupadi who had five husbands may have indirectly encouraged the people to adopt the system of polyandry. During the days of the native rulers the institution was directly helped and encouraged. The rulers used to impose penalties on the partition of properties. In the case of a division of movable property, a one half share was appropriated by the state, and in the case of immovable property, official recognition was denied.

In a polyandrous family the relationship between son and fathers is not difficult to establish. Normally the practice is that all the husbands are recognised as the fathers of each child. The eldest among the husbands is called *Teg-Boba* (elder father) and the others as *Goto* or *Jigich-Boba* (younger fathers). For day to day use the eldest among the living brothers is spoken of as the father of all the children born to a polyandrous wife. In the case of the division of the joint family, the wife names the fathers of her different children.

The arrangement of marital relations in a polyandrous family might appear difficult and strange, but in practice this is not so. The eldest brother, being the head of the family (*karta*) mostly remains out visiting places for earning a livelihood. Earlier there was the trade with Tibet. One amongst the remaining generally looks after the flocks and yet another for trade or other occupation outside home. In practice it is not so often that all the brothers are at home at one time. The tact of the lady plays the major role in the whole system. She may have a weakness for a particular man but in practice she never neglects the others. In this system there is always a surplus of female population, which is left unmarried. Mostly they become nuns or *zomes* and live in monasteries.

Now the massive road building projects and district administration offices and their several agencies, provide people with employment at their doorsteps. This has created a major problem for the institution of polyandry. In former times several brothers in a family were never at home together at any particular time, but now they can all be there, as they get work near their home. This has become a constant source of quarrel among themselves making the life of the lady of the house miserable. Thus the institution of polyandry is gradually on its way out.

For an outsider the morals of Kinner people may appear very lax. In fact in one of the earliest reports an officer of the British regime in India has observed:

> . . . Where there is so little crime, it may be inferred that the morality of the inhabitants is the cause; certain it is there is

less falsehood and theft than in any quarter of Asia. There is a degree of simplicity too amongst these people, and in the interior a modest assurance to be observed that induce an idea of a certain degree of morality existing, but when we take into consideration some of the customs peculiar to them, our belief is shaken. It must be remarked, however, the people consider them more leniently. It may not be so much vice as ignorance. No horror is expressed at the violation of female chastity. Shame hardly exists in some of the remoter states.

The abhorrent custom of polyandry of husbands, the debased state of the sex, all speak the moral depravity of the people to certain extent.[10]

In contrast to the above, the reaction of Kinnauries is very sharp. They resent and defend the institution most vehemently. In their opinion:

> Some superficial and hasty or ignorant or perverted observers of polyandry, as a form of marriage, have at times, sought to read laxity of sexual morals into polyandry. Unquestionably offensive and undesirable as polyandry is by all standards of modern matrimonial sensibilities, it would be illogical, unscientific and thoughtless to mix up a customarily accepted form of marriage with absolute ideas of individual sexual morality. A polyandrous wife has sexual relations with, and emotional attachments to, more men than one because all these men are her husbands by the sanction of custom and it is not merely ethically legitimate but actually a social duty for her to fulfil that attitude towards them all. Such a wife also is true and faithful to say, five husbands (Draupadi had five) is morally superior to a monogamous wife who carries on with even a solitary paramour, to say nothing of five lovers; because this later has played false and faithless to her only husband. . . .[11]

RITUALS

CHILD BIRTH

Pregnancy is not considered an abnormal period here, as elsewhere. An expectant mother leads a normal life. She takes the normal food except during the last one or two months before child birth. She attends to her routine duties except that she is not permitted to do hard labour, or cover long distances, or to go to the houses of relatives other then her parents and that too only if they happen to reside nearby. A few superstitions are in vogue which every expectant mother observes. She is not allowed to gaze at a lunar or solar eclipse. It is believed that if she were to do so the child in her womb is likely to develop certain physical or mental defects. She may not also go to the cremation ground or visit such places which are considered to be haunted. She may not see a dead body. Delivery usually takes place in a separate room other than the main living-cum-kitchen of the family. An experienced elderly woman in the neighbourhood helps in the delivery. A midwife (*dai*) is called *api* in the local dialect and she is respected by the baby like a mother throughout her lifetime. Immediately after delivery hot ghee is given to the mother to eat. *Chuli khali* (hill fruit like apricots or peaches) is boiled in water and the mother and the baby are bathed in it. A daily massage with *chuli* oil is given to the mother and child followed by a hot water bath, during the first fortnight after the delivery. Special care is taken to keep the feet of the mother warm. This is to keep the body supple and strong and nourish the tender skin in the dry and cold climate of the area. Recently rural primary health centres, maternity homes, and a hospital have been provided and people have started taking advantage of these facilities.

During the fortnight after the delivery, the mother is usually served with a balanced diet consisting of ghee, animal-fat, *chuli* oil, barley or *ogla sattus*, wheat *chapaties* with butter *halva*, coconut, and rice. Chillies and sour or tart ingredients are avoided.

The mother and her family are considered ritually unclean for seven days. They observe *begnang*, in which they are not allowed to enter places of worship. The musical instruments and other vessels kept in the temple may also not be touched by them. In case any person from the family happens to violate this rule, a penalty in the form of a *cheema* is imposed. In this *cheema* a he-goat has to be offered to the deity. But the custom of sacrificing a he-goat is fact disappearing and is being substituted by a simple *hawan* (ritual fire). On the seventh day the whole house is cleaned and *gomutra* (cow's urine) and *gangajal* (Ganga water) sprinkled everywhere. The kitchen is given a smearing (coating) of cow dung. The mother is then brought out with the child in a verandah where both of them view the sun. In the verandah seven pebbles are placed over which a mixture of *gomutra* and *gangajal* is sprinkled. The mother then goes round these pebbles seven times and on the completion of each circle she throws away one of the pebbles. In certain cases *hawan* is also performed by a lama on the seventh day. Over this holy fire of *hawan* splinters of *aagru* (a sweet smelling wood), leaves of the *chilgoza* tree, ghee and barley are poured. The lama chants some *mantras*. The ritual purification ceremony is called *chhangshimig*. On this occasion close relatives and members of the clan (*biradari*) depute their representatives to convey blessings to the new born. They bring presents of a few grams of ghee or *chuli* oil and two kilograms of wheat or *ogla* flour or *kanwni* rice for the mother. A feast is arranged for these guests. The offerings are received only from people of the same caste strictly on the basis of give and take. In the plains it is called *bewhar*. The members of the house may now visit the shrines and temples and lead a normal life.

In upper Kinnaur several other ceremonies are observed such as, *khaskis bedhai* and *bose*. In *khaskis*, on the fifteenth day of the delivery, the entire clan (*biradari*) is invited and a big feast with great pomp and show is held. Chamangs and Domangs are also invited, but they are entertained separately and do not enter the houses of the Negis. Champak flowers brought from the lower hills are presented to the father or the grandfather of the child. The *api* is given food daily during her attendance on the mother for eight days and thereafter she is given clothes and some cash. In recognition of her services she is invited off and on and entertained for three years at least. In *Bedhai* congratulations are offered by Chamangs and Domangs and on *Chait Sankranti* day, i. e. in the first month of the year. Musical instruments are worshipped by offerings of wine and wheat or barley bread. After worship, the offerings are distributed among those present on the occasion. The *bose* ceremony is held between the child's first and third years. The local deity is invited to bestow his blessings. In Kanam village the Dabla deity is requested to reveal his wishes which are expressed through the *maligorkh* (medium worshipper). The *grokh* goes into a trance and demands offerings of wine, wheat bread, and *munbhog* (offering). Some coins and a *gagro* (piece of cloth) are offered to the diety. The host will then present all the aforesaid articles to the *Kardars*. Thereafter, the invitees are served wine in small *Kargvols* wooden bowls with *Musarho* (preparation). Women are served *chhang*, which is prepared by fermenting barley and which contains very little alcohol. *Chiltas*, *palis* and barley *sattu* are served after the wine. The kith and kin then present *dagloo* (bangles) and clothes to the child. In addition to this, some bring barley *sattu*, wheat flour, butter and *chuli* oil. After the feast is over the *baskisa* play on their instruments and the relatives, both men and women take part in the dance that follows. The father of the child leads the dancing party and in a semi-circle all move backwards and forwards to the tunes of the local bands. Songs are sung befitting the occasion. This merrymaking goes on for several hours and often continues through out the night.

KRACHOGMIG (MUNDAN CEREMONY)

When the child attains the age of one or two years, his or her *mundan* (shaving of the head) ceremony takes place on an auspicious day in consultation with a Lama. This is always done on full moon day or during the bright half of the month, i.e. *shukla-paksha*. Every odd month is considered auspicious. On the appointed day the maternal uncle, or in his absence some other person whose mother and father are alive, cuts the hair of the child with a pair of scissors. In the plains the barber shaves the head with a razor. The hairs are then buried. In the plains the hair is collected and thrown in to some holy river. Some friends and near relatives are given a feast also on this day. There is no difference in the ceremonies performed on the birth of a boy or girl. In fact the birth of a girl is not taken with a heavy heart, despite the large number of nuns or *zomos* in the area.

In case a child happens to cut the upper tooth first then this is considered a bad omen for the mother's parents and brothers. To ward off the evil effect, the mother or her husband would send a message regarding this to the maternal uncle. Thereafter, the child's maternal uncle would bring a few clothes for the child and before entering the house would drop these clothes through the *Dusrang* (outlet for smoke) and the mother would collect them and put them on the child immediately. The maternal uncle would then enter the house through the main door. This is believed to ward off the evil effect attendant on the cutting of the upper tooth first. In the plains the maternal uncle gives some money and a metal cup to the child.

Except for the ceremonies of *bedhai* and *bose* observed in the upper reaches of Kinnaur, all others in some form or the other are performed in the rural areas of the plains. The ceremonies of *chathi* (sixth day) and *barahwan* (twelfth day) and *mundan* (head shaving) are just the same. Also, the superstition about the cutting of the upper tooth first. The difference lies only in some of the materials used for worship etc. For instance in the plains seven lumps of earth are picked up, in place of stones, in place of *chuli* oil, mustard oil is used; in place of wine water is used; in place of *sattu* barley is used, and mango wood is used for *hawan* purposes etc.

MARRIAGE

Marriage rituals in Hangrang tahsil in upper Kinnaur differ from the rest of the region. They have nothing to do with either love or affection. They are mere contracts of convenience and a matter of a bargain by the parents of the young.

Rehja, as marriage is popularly called here is settled much earlier while boys and girls remain very young. Settlement of the marriage is done by offering a piece of cloth (*khatak*) and a bottle of liquor by the father of the bridegroom. If these offerings are accepted by the father of the bride, the marriage is considered as settled. When both boy and girl attain the marriageable age, the bridegroom's father alongwith some other persons goes to the prospective bride's house and agrees upon the day of the marriage. At the same time, they also agree upon how much money the bridegroom's side will give to the bride's family. This money varies from rupees fifty to one thousand, depending upon the capacity and the status of the family concerned. This money is paid before the date of the marriage and is used in the purchase of clothes and ornaments for the bride. This expenditure is accounted and taken note of. In the event of the wife leaving her husband and going with another man, this money has to be returned.

The money transaction in the above type of marriage seems to be extraordinary. Can it amount to a bride price or *reet* money as is the practice in several other parts of Himachal Pradesh? But the people of Kinnaur do not admit it as *reet* or bride money. In their opinon:

> The custom of *reet*, once pervalent in certain parts of Himachal Pradesh, has never been in force or in practice in Kinnaur. It would be a gross misreading of any custom of Kinnaur to call it, or liken it to *reet*. There never has been any custom of the bridegroom's father or guardian or anybody else paying any bride money. The word *dhari* alleged by some as the name of this supposed bride price is an utter stranger to the main Kinnauri language as well as to the minor dialects and as much a creation of a pure fiction as the very notion that any such payment passes between the parties to a matrimonial alliance in Kinnaur.[12]

The standard and legal marriage is known as *Janetang* or *Janekang* in Kinnaur. In the details of the rites, there are some local variations in different parts but the broad outlines are almost alike everywhere. In this form of marriage a suitable bride is looked for by the parents of boy and then the negotiations are opened with the girl's parents generally through relatives. If the bride's parents seem agreeable to the proposal, two persons called *majomi* (go-between) carry a bottle of liquor and some cash, generally five rupees, as token gifts and go to the girl's parents to settle the marriage. Acceptance of the proposal is followed by an offering to the family deity or the village deity and, in some places, a token item of ornament is left behind by the middleman. In some areas, the offering to the family deity consists simply of a bottle of liquor topped by a lump of butter. The butter is portioned off and applied to the forehead of all present and the liquor distributed in spoonfuls to those who drink, as token of the deity's blessings. After this a marriage date is fixed in consultation with a lama or the village priest. On the appointed day, the bridegroom goes to the bride's place accompanied by a small party of half a dozen or so along with the village band. On reaching the bride's house, the party stays there for two feasts (evening and morning) held in their honour.

Traditionally, the bride is not supposed to be officially in the know of the happy event till the groom and party arrive. After the arrival of the party, she is groomed up for the occasion. Merrymaking during the night includes much singing and dancing and the following day the bridegroom returns home with the bride and the bridal procession comprises anything up to a hundred or even more persons. On reaching the groom's village the party assembles at the temple for some time. The ladies belonging to the bride's village but married in the groom's village go out to welcome the bride and take with them wine and other eatables for the guests. A *puja* is first performed with these things to placate the evil spirits supposed to have come with the party. Later, these offerings are eaten by the members of the party and there is rejoicing. After this they go to the groom's house. Those foremost in the reception of the bridal party include the bridegroom's sister or mother holding a container full of grain (barley) with a lamp (*deewa*) atop the grain heap. Songs, wedding songs chiefly, and dances are the chief features of the festivity that ensues and lasts through the whole night. The principal symbolic ritual is called *oo-pagey*. In this the groom and his brothers, in case of a polyandrous marriage (whereas in non-polyandrous a single groom) sit in a row and a *pagri* (turban) is wrapped round their heads by the groom's maternal uncle. Garlands of *chilgoza*, *chuli* and walnuts are draped round their necks. The bride also offers these garlands to them. Where there is no turban ceremony, as a custom to symbolize the union, sipping from the same cup by the bride and the groom takes the place of this ceremony. In a polyandrous wedding all the brothers sharing the alliance sip, and so does, last of all, the bride, from the same common cup. Presents, whether in cash or in kind are made by friends and relatives to both bride and bridegroom. In some parts of the district the bridegroom does not go to bring the bride and she comes to the groom's house with the middlemen whose number in such cases, runs into quite a small party. In these areas, such an induction of

5. A typical house in Kinnaur

6. Kalpa Bazar

7. Village headman in Buddhist monestry

(*Facing page 32*)

8. Kinnaur school teacher with a lady

9. Kinnaur Kailash

10. Kinnaur Kailash

(*Facing page 33*)

the bride into the bridegroom's family, with certain brief ceremonies, is considered good enough to serve the purpose of a marriage and the full dress wedding ceremony is performed later at a suitable date.

After the marriage is over, the groom's father sets aside some piece of land for the bride and this is confirmed in writing on a plain or stamped paper. The middleman stands as witness and the paper remains with the girl's father or maternal uncle. Similarly, a surety bond is given by the bride's father to the effect that the said document shall become inoperative in case of a divorce. All the utensils are now given and a list is prepared after weighing each one of them. In the event of a divorce these utensils or the cost thereof has to be reimbursed by the groom's father. This custom is called *bandobust* ceremony. Can the non-tribal civilized world boast of any such arrangement for their daughters in case of distress? It is perhaps a unique custom that is found in Kinnaur tribal society.

After the rituals all the near relatives of the groom, who have been invited on the occasion give some money as gifts to the groom's father, and he alone is the master of this collection. This in the local language is called *baildayang*.

Similarly, every man from the bride's side gives a cash gift to the bride. This ceremony is known as *udanang*. An account of this collection is kept on a piece of paper by the bride's father or maternal uncle. The money can, however, be utilized by the girl's father for three years without any interest and thereafter an interest at the rate of 12½ per cent is charged. Ultimately, the bride utilizes this money in buying a piece of land or ornaments as she thinks proper.

At the end, the bride will garland each of the members of the marriage party in the *chuli* or *chilgoza* thus bidding farewell to them, weeping bitterly and embracing them. This ceremony is called *borshimik*. For the first time, the bride stays for a brief spell and then returns to her father before the end of a month.

The *janetang* or *janekang* form of marriage is considered as a wedding. The other forms of marriage in this area are: *dumtangshis* or *bennang-hachis* or the *jushis* (love marriage);

dubdhub or *khuchis* (marriage by capture); and the *har* (enticing away someone's wife).

Love marriage is called *dumtangshis*. In this form of marriage the boy takes his beloved to his house secretly. In marriage by capture, i.e. *dubdhub* or *khuchis*, the boy with the help of a few friends forcibly carries away the girl of the choice to his house. In both of these cases, the boy's father deputes two middlemen to the girl's parents requesting for their pardon for the outrage committed by his son. Initially, the girl's parents show great resentment and their attitude is thereatening, but the middlemen finally succeed in bringing them round. The girl's father asks them to make amends in terms of money as a sort of atonement for this disgraceful act by the younger one. The money so demanded is called *izat* money and it varies. This money has to be deposited at once by the middlemen. Sometimes the girl's father also demands that the wedding type of marriage ceremony be performed. In that case the boy's father would request for the exemption from such a performance by expressing his inability to bear the expenditure and the exemption is always given. Within a month the girl is sent back to her house—on some auspicious date the boy accompanied by the middlemen and two ladies escort the girl back to his house. On this occasion she is given new clothes and a few ornamenis which they can afford. The boy offers *chilgoza* and *chuli* garlands as a mark of respect to his mother-in-law and other elderly women in the family. A few rupees according to one's capacity are also given. After that the groom returns with his men leaving behind the bride. On a suitable date thereafter her father and brother take her to her husband. All the relatives collect at the girl's house on this occasion and absolve the girl of the disgrace of being forcibly carried away or of elopement. The girl then goes to her husband and her escort returns after a few days stay.

In *har* or *hari* a woman runs away or elopes with a man other than her husband. In this case the husband will send a man to the new husband and demand a refund of the expenditure incurred by him during his marriage. He will have to be repaid this amount which is called *hari* and is mutually decided upon. This type of marriage is not very different from *reet* which was prevalent in some other parts of Himachal Pradesh and which permitted a woman to take another husband who would pay to the former husband the amount incurred by him on his marriage with the woman concerned.

The dowry system has never existed in this area. As noted earlier, gifts and presents are customarily given to the bride by her parents and relatives and *udanang* are regarded as an exclusive property of the woman. At the time of marriage, the bridegroom is given a turban, a *khakharis* a *nangeh* (a dish) or a *batich* (cup) and a *nang* (platter) and sometimes he is given some garments. These gifts cannot be regarded as dowry.

Divorce is also permissible among all castes and in case a woman wants a divorce she usually does it through her parents. The divorce proceedings are very simple. On an appointed date the parents of the woman and her husband or husbands gather at a place. They then settle the accounts. The utensils, ornaments and cash given to the woman at the time of marriage by the parents have to be returned by the husband or the husbands. Finally, a twig is placed between the couple and they break this twig, which symbolizes the breaking of the marriage. She becomes free to look for another husband. Widow remarriages are also allowed.

The description and defence of the marriage customs of Kinnaur by them is as follows:

> The straight course in love marriage, is for the parents or guardians, as the case may be, on both sides, to themselves sense, like sensible elders, or to be made to realize tactfully by well wishers, that boy and girl have chosen each other and that wisdom and decency lie in formalizing the romance into a conventional marriage. Thereafter the routine is that the standard procedure of an arranged marriage is fully followed. An alternative is for the elders on both sides, or on the girl's side, to feign ignorance of any intention of the suitor to whisk away his beloved, and thus to connive at such action by the lover and his friends. This alternative is resorted to when a balance of convenience dictates this short-cut with the tacit consent of the girl's elders and the open participation or implicit concurrence of the boy's people. She is carried off in two other circumstances. When, in a love marriage, the parents or the guardians on both sides or either side do not favour the alliance, the lover and the beloved take the

social law into their own hands and the former acts by carrying off the latter. At times it takes quite an organisation to do this, because the girl's elders might physically resist the attempt. When the intended bride has to be carried off without even the tacit consent of her parents or guardian, then the proceeding smacks in 'spirit' of the primitive origins in marriage by capture, though softened in the actual execution by civilized touches. This custom has very rarely been practised for long now and may be taken as practically abandoned. When there is whisking away or 'carrying off', whatever be the background the marriage is invariably, sought to be regularized by securing the approval of the parents or the guardians and, as a rule, the ceremonial seals of social approval (be it simple or elaborate according as the elders from both sides agree upon) follows in due course. At times if local concepts of social dignity so require, a perpetrator of marriage by capture or by carrying off may be penalised with fine. Elopement is not a recognized custom. Marriage by intrusion is almost extinct now, though, in times bygone, it used to be a rare custom to meet extraordinary circumstances when a lover, having enticed his beloved into physical intimacy by a promise of marriage, backed out of the commitment and had to be forced by the pressure of social conscience to recover his dignity and accept the girl as his wife. At times this involved the girl's entering the husband's home uninvited and unbidden. Eventually, even in such contingencies society usually found ways and means to regularize the alliance.

. . . There is nothing whatever in the marriage customs, ceremonies and traditions of Kinnaur which might reasonably be interpreted by anyone as sale and purchase of and a bidding for, a woman whether it is her first-marriage or remarriage.[13]

DEATH

In the event of death, the body is cremated. Infants up to the age of two years who happen to die are buried. In the case of children above two years and below five years the bodies are thrown into the river. Those who die of leprosy or small pox are either thrown into the river or buried. It is likely that in ancient times a dead body was kept on the hill top to be eaten by animals and birds.

Normally a dying person would be lain on the floor as no

cots are used in the area, but in case such a person happens to be on a cot he or she is brought down on the floor just before the end. The place where death takes place in either besmeared with cowdung or some soil from a sacred place is placed under the bedding. Just before the death *panchratna* (alloy of gold, silver, copper, brass, and iron) is kept in the mouth. A lama is often called to stand by and to recite some *mantras* and to pray for the eternal peace of the soul. The dead body is bathed in warm water. On the spot where death takes place a brass lamp is lit with *chuli* or mustard oil, this continues to burn for seven days continuously. Besides the lamp a pot full of barley is also placed.

The funeral procession is accompanied by a band playing sad tunes. The man leading procession carries an urn with live coals. Barley is scattered at frequent intervals all along the last journey. As the procession starts, some money in loose change is thrown over the bandmen which they pick up. One man from each household in the village accompanies the corpse for collecting wood and other materials for the pyre. A man from the Koli family carries some Kail or *neoza* torchwood and an axe to the cremation ground. If death occurs in a Koli family they arrange this between themselves. The man who performs this is called *Binanes*.

The pyre is made of wood and the body is placed over it. The son or the nearest relative then takes some water in the hands and goes round the pyre once and pours the water on the head of the body. He then lights the pyre from the direction in which the head is placed. Everyone then returns home except for the nearest relatives who remain there till the body is completely burnt. The ashes are carried mostly to Hardwar for immersion in the holy Ganga. A lamp is kept burning continuously near the urn containing the ashes till the immersion, either in the Ganga or the Sutlej. The near relations shave their heads as a mark of respect to the dead. The mourning continues for eight days during which period the family members do not eat meat, oil, fried articles of food, onion, garlic or turmeric, nor do they season with *tudka* or *chhuk* (spices fried in heated oil) the cooked pulses or vegetables. During these days early in the morning cows are given some food. Some bits are also scattered for the crows. The food given to them is supposed to reach the departed soul. On the eighth day the family members purify themselves and the house is also cleaned both actually and ritually. In the evening they get together and a feast is held in which no wood taboos are observed. In the local language this ritual is called *kolyash-inmg*. On the fifteenth day a general feast is held for the villagers in the temple where only *chiltas* are distributed and these are taken home. But the lama and the persons who were present at the cremation right to the end are fed with good food. The last ritual after a death is known as *Aating*. It is performed on the thirteenth day festival following the day of death. On this day all the relatives and the villagers are given a grand dinner in which *poltus* are distributed. Musical instruments are also played on this day and flowers are exchanged by the relatives. This marks the end of the funerary rites.

There are no separate customs to be observed by a widow except that she removes all the ornaments worn on her person. She is free to take another husband after the *aating* ceremony. In the case of a polyandrous family, a women never becomes a widow unless by ill-luck she loses all her husbands.

Except the wedding and Hari type of marriages, all other forms with variations are also found in the plains. For instance love marriages are not rare these days nor elopement and court marriages are unknown. The funeral customs of these people are remarkably similar to the customs and rituals observed in the rural areas of the plains. Except that in place of Koli a barber accompanies the bodies to collect wood etc. Shaving of the head amongst the near relatives is very common, so also ceremonies like *suddhi*, *duswan* and *terhi* (thirteenth day) etc. are performed in which Brahmins are fed and clothes and other things used by the dead are give away *dan* or as gifts.

COMMUNITY LIFE—FAIRS AND FESTIVALS

The only means of amusement and recreation for the villagers are fairs, festivals and community dances. The fairs are held on festival days and are accompanied by dancing all round the year in different parts of Kinnaur. During these fairs and festivals, they drink and dance in joyful mood casting all the worries of life aside for some time. At one of the festivals, they often dress themselves in queer costumes like masks with horns and fancy dress and amuse themselves, but no magico-religious importance is attached to these masks. The women dancers sing songs which invariably consist of some old or contemporary local legends set to simple tunes. Musicians, mostly Harijans, with drums, wind-instruments and cymbals accompany the dancers as they circle, arms entwined, with slow, rhythmic steps round their village deity, seated in state in the temple courtyard. They make a most pleasant picture in their best, festive attire, bright faces shining with simple pleasure as they dance and sing under the pure blue skies, either under a bright sun or in a strong breeze.

There are several types of dances performed in different parts of the district. These are classified in two different broad categories of *kayang* and *bonyungabu chashimis*. These divisions depend upon movements of steps, style of holding hands, variety of songs sung and music played during the dance. These dances are further classified into different finer categories. For instance there are some six varieties of *kayang*, such as *babaskayang*, *pulashom*, *somahelang*, *tegsowang*, *bang-paroshunig* and *thungru*. The two varieties, *chama* and *shonma* dances are performed in upper Kinnaur mostly.

To begin with the orchestra starts playing tunes calculated to attract the dancers. These rhythms are finally changed to *Kayang* or *Bakayang*. Thereafter men and women join to form an arc or circle. Normally the elderly men remain near the leader. Similarly in their own section of the circle elderly women lead the ladies. In front of the leader within the arc or circle, move the musicians playing their instruments. Any two women who are conversant initiate the song first. This is called *Balgithang*. The lines are repeated in chorus by all the dancers. The repetition is called *Jagithang*. In this fashion the dance continues for hours together, the song and the leader keep changing at frequent intervals. These dances go on sometimes till the early hours of the morning.

Fairs in Kinnaur district do not always signify commercial or trade activities of the region. They are organized generally on the festivals.

> The most interesting feature of the majority of festivals is the zest and the genuine love of flowers—which people generally love. The effort they make to collect them and the joy they express in dancing and singing, on the collection of flowers from high hill tops are indicative of that love. After a difficult ascent to the meadows and after even spending a night in the caves, when eventually the bonquets are prepared, the men present them to their ladies and to their deities alike.[14]

Different regions of Sangla, Kalpa, Puh, and Hangrang have their different festivals and fairs, but a few festivals are celebrated all over the district such as *Dakhraini*, *Beech*, or *Baisakhi*, *Oo-Khayang*, *Sazo*, *Phagul* and *Jagro* etc. A description of some festivals is given here to give an idea of the gaiety of the people of this region.

BAISAKHI

Baisakhi or *Beesh* is the first festival of every Vikram Sambat year and is celebrated on the 1st *day* of *the month of Baisakh* (See Appendix). Villagers prepare good food like *polhu*, *halwa* and *koyashid.* The image of the goddess is brought out of the temple and a fair is held in the courtyard of the temple. It is an occasion to get together and to dance and drink. Harijans dance separately about twenty yards away, the musical instruments, however, are not played for them. This marks the end of the winter season also. New woollen clothes are worn from the wool spun during the winter.

DAKHRAINI

Dakhraini festival is celebrated on the 1st of *Sawan*, in which a feast is served. *Chhack* milk (curd) and butter are specially enjoyed. No animals are sacrified however. The deity is brought out and the villagers dance before her. Zongor and Lhoskar flowers are brought from the Kanda peak and their garlands are offered to the goddess. After this these flowers are distributed among the villagers. One or two members from the family where a death might have occurred before this festival go to the peak of the hill and offer some food and fruits to the shepherds in memory of the departed soul. A white flag on which some Buddhist mantras are written is fixed there as a sort of prayer for the peace to the departed soul.

FLAICH OR OO-KHAYANG

Flaich or *Oo-khayang* is essentially a festival of flowers celebrated during the months of *Asauj*. This festival is celebrated throughout the Kinnaur district on different dates. Generally people celebrate it on the hill peaks near their villages. During this festival, the villagers leave their houses and move to the said place leaving only a few old persons behind. The village deity led by a band is carried to the place in procession. At least one member from every household must accompany the deity. One he-goat is sacrificed there, immediately on arrival of the deity of Kali, the spirit of the peaks, and wine too is offered to it. The next morning again at the time of *puja* a ram or a he-goat is sacrificed. A fair is held throughout the day. Flowers of *shulloo* which have been brought from the peaks for the purpose, are woven into garlands. At the end of the fair these garlands are offered to the deity. It is believed that with the flowers the spirit of Kali comes to the fair. This spirit has to be given a farewell and the *grokch* (*Pujari*) goes up a few furlongs and sacrifices a he-goat and bids farewell to the spirit. Immediately after that the people accompany the goddess back to the village adorned with flowers. The people sing and dance on the return journey.

SAZO

The *Sazo* festival is observed on the 1st of *Magh* every year. On this day the people take their bath in the natural springs and a few even go to the Sutlej river for bathing if they happen to live near the river. One or two elderly ladies in each household get up at the midnight and take their bath. After this a feast has to be prepared for them before dawn. *Poltus*, rice, pulses, vegetables, meat, *halwa*, *chilta* and *pug* are the principal dishes, prepared on this occasion. In the morning the family god is worshipped with the food except the meat. The hearth is also worshipped. During the night guns are also fired to ward off evil spirits. Near noon time the deity is brought out and worshipped with wine and *halwa*. A folk-dance is held. Thereafter the deity is believed to have gone to

Kinner Kailash. It is believed that during the month of *Magh* (January-February) all the gods and goddesses of Kinnaur rest for a couple of weeks during which period they attend a meeting presided over by Lord Shiva on a mountain peak, facing Kothi village known as Kinner or Banasur's Kailash. This peak has been named after Banasur the mythological demon King, who once ruled over the area and did penance on this peak to earn a boon from Lord Shiva. It is believed that the conference of deities ends with the bestowal of the boon by Lord Shiva. Thus if a particular deity returns to its land with the boon of prosperity, the land will yield bumper crops and other blessings that year, otherwise there will be drought and other afflictions. It is through the oracle that villagers learn about these things and prepare themselves for the coming events. When the deity is supposed to have gone to Kinner Kailash all the adornments are removed and the temple is closed for the period. This is no longer believed in very religiously. The people are now gradually developing a scientific outlook about such beliefs.

PHAGUL OR SUSKAR

Suskar festival is celebrated mostly in the month of *Phalgun*. In this festival the spirit of Kanda (Peaks) called Kali is mainly worshipped which lasts about a fortnight and is celebrated all over Kinnaur. Each day of the festival is called by a different name and several peculiar functions are held each day. On the last day a feast is prepared and people worship Kali on the roof of the house and then partake of the food. It is believed that if the functions and festivities are celebrated with full zeal, Kali the spirit feels happy and blesses the villagers with prosperity and plenty in the coming year.

DEEVAL

This is celebrated on *Amawas* (new moon) of Diwali. This has, however, nothing in common with the Diwali (Festival of lights) of the rest of the country. On this day the villagers clean their houses and plaster the walls and the floors with cowdung. At night some place in the compound of the house is plastered with cow-dung. Some splinters of torchwood are placed crossing each other. On the four corners of the plastered area is placed a mixture of *chilgozas*, walnuts, *chullies* and roasted barley, wheat and *chollaie* called *pug* in local dialect. The torchwood is lit and some of the articles are fed to the fire. *Puja* is performed with *poltu*, wine and *dhoop*. At the time of this *puja* the head of the household or the person performing it chants the *mantras*. *Deeval Patti Vahu* and offers some *zongor* flowers to the fire. Good food like *poltu*, *halwa* and wine are taken with great relish. The Goddess is also brought out and a fair is held.

SHIVRATRI

This festival is specially celebrated by Baddhis. Houses are cleaned. One man from each family is sent to bath in the Sutlej river and on his way back he brings *Zangazang Narkasang*, flowers and leaves of *Karmanyang*. At midnight a thick *roti*, rice, *rulpole*, *mast*, *halwa* and *poltu* are prepared. A *shivalinga* (phallic symbol) is prepared with the flowers and leaves. The *linga* is hung from the roof. A temple is made on the floor with flour.

The entire family keeps awake during the night and local songs are sung in honour of Lord Shiva and the food is placed round the temple made of flour. This is done by the person who has come from the river after taking his bath and has brought the flowers and the leaves. He avoids touching other members of his family before the puja as he is ritually pure. A lamp of ghee or *chuli* oil is also lit. He then worships the lord with incense and *kungu* etc. The person performing the puja fasts throughout the day.

ART AND CRAFT

There are no modern sophisticated artists in the area nor is there anyone who can be called an abstract artist, yet the Tibetan style of Buddhist painting has had some seven generations of artists in the Nesang Village of Kinnaur. A number of other villages have also been producing artists of this style, though the ancestry does not go far back as in Nesang. Mostly, the artists had learnt their art in Tibet in bygone days. Only a few have been trained locally. The original tribal paintings have been plain and rudimentary.

Kinnaur has known artefacts and handicrafts that promise

great attraction as antiques and decoration pieces no less than in some cases, as utilitarian objects. It may be an ornament, a matchlock, an inkhorn, a tender box, a tea cup, a general purposes cup, an incense-burner, an oil lamp, a tea pot, a piece of furniture, a statute, an idol etc. The ornamental designs on the fabric or borders of certain garments are highly picturesque. In fact every village has some craftsman or the other, who prepares several objects and utilitarian things like weaving of *dhori*, *chhauli* shawls and other woollen clothes used in the villages. Blacksmiths and silversmiths are the craftsmen who make agricultural tools, locks and utensils and the jewellery commonly used by ladies of the region. Potters and basket makers, manufacture earthen pots for household use and baskets etc. for the purposes of agriculture and other use. Woollen shoe makers are available in almost all the villages, as they are indispensable in this cold region.

The village pattern in this area is almost similar to the villages in the plains where all the people along with craftsmen most often needed in agriculture and household live together. They are mostly barber (*nai*), washerman (*dhobi*), waterman (*kahar*), potter (*kumhar*), carpenter (*badhai or tarkhan*), ironsmith (*lohar*), goldsmith or silversmith (*sunar*), weaver (*kori*), cobbler (*chamar*), basket makers, and agriculture labourers etc. These craftsmen get a share from every household in kind commonly known as *jewra*.

REFERENCES

1. Captain Alexander Gerard, *An Account of Koonawar in the Himalayas, etc. etc.* (London, 1841), p. 29; Several European and Indian travallers of nineteenth and early twentieth century have praised these qualities of Kinnauri people. The reference can be made to the following works for detailed description of the people; James Baillie Fraser, *Journal of a Four Through part of the Snowy Ranges of the Himalayan Mountains and to the Sources of the River Jumna and Ganges* (London, 1820); S.R. Johari, *Our Border Lands* (Lucknow, 1960); Gordon Cumming, *In the Himalaya and on the Indian Plains* (London, 1886); Van Der Sleen, *Four Months Camping in the Himalayas* (London, 1929); Macro Polis; *Peaks and Lamas* (London, 1939); A.H. Franke, *Antiquities of Indian Tibet* (Calcutta, 1914, 1925), etc.

2. Lloyd William and Capt. Alexander Gerard, *Narrative of a Journey from Cawnpoor to the Boorendo Pass in the Himalaya Mountains, etc.* (London, 1840), Vol. II. p. 206.
3. For Languages the following works can be referred to: Rahul Sankrityayan, *Kinner Desha* (Allahabad, 1956), pp. 376-90; T.S. Negi, *Scheduled Tribes of Himachal Pradesh* (Simla, 1976), p. 19; S.S. Shashi, *Himachal: Nature's Peaceful Paradise* (Delhi, 1971), p. 109; M.D. Mamgain, *Himachal Pradesh District Gazetteers: Kinnaur* (Simla, 1971), pp. 69-76; G.A. Grierson, *Linguistic Survey of India* (Delhi, 1968), Reprint, Vol. IX, Part IV, pp. 613-46.
4. *A Glossary of the Tribes and Castes* (Lahore, 1911), Vol. II pp. 446-47. For Kanets in general see from p. 456.
5. Census of India 1961, *Housing and Establishments of Himachal Pradesh*, Vol. XX; Part IV. pp. 339-54.
6. Mamgain, n. 3, pp. 114-15.
7. Negi, n. 3, p. 39.
8. Edward Thornton, *A Gazetteer of the Territories under the Government of the East India Company and the Native States in the Continent of India*, p. 518.
9. Fraser, n. 1, pp. 255-56.
10. Captain C.F. Kennedy, Assistant Deputy Superintendent, Sikh and Hill States to Lt. W. Murray, Deputy Superintendent, Sikh and Hill States, 6 July 1824. *Punjab Government Record: Delhi Residency and Agency, 1807-1857* (Lahore, 1911), Vol. I, pp. 255-99, para 25.
11. Negi, n. 3, pp. 29-30. on Polyandry; See Dr. Y.S. Parmar, *Polyandry in Himalaya* (Delhi, 1975) and Dr. D.N. Majumdar, *Himalayan Polyandry* (Bombay, 1962); R.N. Saxena, *Social Economy of a Polyandrous People* (Bombay, 1954).
12. Negi, n. 3, p. 33.
13. Negi, n. 3, pp. 33-35.
14. Mamgain, n. 3, p. 116.

3

Historical Perspectives

TRADITION TELLS us that the region now known as Kinnaur district had been the place of abode of the legendary Pandavas for a year during their period of hiding (*Agyatwas*). Perhaps, that is why some object or other, either a tank, a temple or a gate is associated with them even now. Some people trace the origin of polyandry to the legend that Draupadi had five husbands, i.e. the five Pandava brothers.

Whatever may be the origin, there is no doubt that the Kinner have deep roots in Indian mythology, legends and literature, and are considered as a distinct race, somewhere between human beings and gods. Perhaps with the sole exception of the *Rig-Veda*, they are referred to in all ancient literature in some form or the other, though there is no unanimity in literature on the question of their origin.

According to *Rajtaringini* the Kinner were born from the shadow of Brahma,[1] the *Amarkosha* mentions Vidyahara, Apsara, Yaksha, Rakshasa, Gandharva, Kinner, Pisacha, Guhyaka, Siddha, and Bhuta having the same origin as the Devas.[2] Excepting Rakshasa, Pisacha, and Bhut, all the remaining races are supposed to be well disposed towards the good.[3] According to the *Matsya Puran*, Kashyap Rishi had thirteen wives. Among them Muni and Arista gave birth to the class of *munis* and *apsaras* and the race of Kinners and Gandharvas respectively.[4] Another tradition says that out of eight *Rishis*-progenitors, the last three, Pulastya, Pulaha, and Kratu produced no true Brahmins. It is said that the offspring of Pulastyas were Rakshasas, Vanaras, Kinners, and Yakshas, and the offspring of Pulahas were Kimpurushas and Pishachas.[5] The *Vishnu Purana* says that one Priyavarta Raja had seven sons. He divided his kingdom among them, and Jambudwip went to Agridhara. Agnidhara had nine sons, one of them was Kimpurusha. When Agnidhara divided his kingdom Jambudwip amongst his sons, Kimpurusha got Hemkut region.[6] In the *Ramayana* the Kinners are mentioned along with Devas, Gandharvas, Charanas, Siddhas, and Apsaras. 'The Kinners were an effeminate race always represented as indulging in amorous sports'.[7] Manu, the great law giver mentioned that the sons of Atri are said to be the fathers of Daityas, Danavas, Yakshas, Gandharvas, Uragas, Rakshasas, Suparnas, and Kinners.[8]

Kinners are mentioned both with Devas and Rakshasas alike. In the *Ramayana*, Devas, Gandharvas, Charanas, Siddhas, and Apsaras are mentioned with Kinners. In *Amar-kosha*, the Vidyadhara, the Apsara, the Yakshas, the Gandharvas, the Pishacha, the Guhyaka, the Siddha, and the Bhutas have been mentioned with Kinners. In the *Matsya Purana*, the Kinner are associated with Yakshas and Gandharvas. Kalidas has mentioned Kirats, Utsavasanketas, Vidyadharas, Kimpurushas, Punyajanas, Yakshas, Siddhas, and Ganas with the Kinner.[9] In the *Vishnu Purana* they are associated with Gandharva, Yaksha, Rakshasa, Daitya, and Danavas as the inhabitants of the mountains and as singers, dancers and merrymaking people.[10] *Shiva Purana* mentions that Yakshas, Rakshasas, Gandharvas, Charanas, Siddhas. Daityas, Danavas, Sesa and other serpents, Garuda and other birds, all the Manu Prajapati, Kinners, men etc. have worshipped the wish yielding phallic image with great devotion and have achieved their desire surging in their heart of hearts.[11] The *Mahabharata* mentions Kinners along with Gandharvas, Yakshas, Siddhas, and Kimpurushas. In the *Mahabharata* Kinner and Kimpurusha are two different people.[12] The *Adi-Purana* mentions that in the great heights there was a town called Kinnergeet, where Kinnaries used to sing songs always.[13]

Several tribes mentioned in these texts along with Kinners must have been the inhabitants of this earth, most probably in the mountain valleys. Kinneras are uniformly mentioned as great singers and dancers. They may have been the original actors, dancers and singers of this earth.

Buddhist and Jain literature have mentioned the Kinneras time and again. Gautama Buddha had preached his doctrines to Yakshas, Gandharvas, Asuras, Garudas, Kinneras and Mahoragas and perhaps they all had similar faith in them. In order to emphasize the concept of celibacy, Buddhist moral code stressed on the Bhikhshus that if they had intercourse with Kinnaries then they will be guilty of highest offence, i.e. *Parajika*.[14] There are several stories in *Takkariya Jataka* regarding Kinneras and their physical forms etc.[15] Thus, besides Buddhist and Jain literature, Kinneras find mention in almost every religious text of Hindu scriptures, hymns and *mantras* from time immemorial. These legendary, mythological, and literary descriptions of Kinneras may be facts, but if they are not, they most certainly indicate a very highly creative and inventive genius of the authors of these works.

Kinneras have found a prominent place in Indian and Central Asian art. Countless sculptures have immortalized them. The whole of the ceiling of cave one at Ajanta has done them a great honour.[16] Stories of Jatakas have perhaps contributed in the depiction of several physical forms of Kinneras, such as horse-headed and half-bird, half-human figures. The Kinnari and Yakshi motifs have also found place in the Central Asian art. Central Asians have also depicted the Kinners and Kinnaries in their works.[17]

The land of Kinneras at one time was supposed to lie between the sources of the Ganga in the east and the Chandrabhaga river in the west.[18] It is also said that this land was spread between the rivers Jamuna and Sutlej.[19] To the north of it was the southern part of Kashmir, probably Zaskar range, which was forming the northern boundary, and the southern boundary was Dhaula Dhar.[20] This region may have included Rampur and Sarahan, the Shonitpur of the *Puranas*. It was the capital of famous Kinner King Banasur, who according to a legend like Sagara of *Raghuvansa* brought the Sutlej river from the Lake Mansarovar. It is probable that the legendary mythological Kinnerdesh was situated between 3700 and 4,600 metres in the Antargiri or middle Himalayan range between Ganga and Chandrabhaga rivers. Its northern portion was Hemkut mountain which represents the modern Ladakh-Kailash-Trans-Himalayan chain.[21] It is mentioned along with Gandha Madan Mountain in *Amarakosha*. According to a tradition Hemkuta Mountain was the dwelling place of Kinneras, Gandharvas and Siddhas.[22] This land of Kinneras was also called 'Kimpurusha Khand' in the *Puranas*. In this sacred Khanda was a stream of honey that flowed from the *Palasha* tree, and the residents whose complexion was like that of a heated gold, used to drink it and enjoy. It was probably these people who were mentioned as present in *Rajyoga* of Yudhisthira in the *Mahabharata*. It was probably this land where Arjuna during his conquest reached first; the Hatakdesh of the Guhyayas and thereafter the Mansa Lake. Again perhaps it was this very land where legendary Raghu went after defeating the Kiratas, Parvatiyas, and Utsavasanketas.[23] The modern Kinnaur district in Himachal Pradesh is the remnant of the old legendary and mythological land of Kinnerdesh. There must have been pressure from different sides on these people. It is probable the Vedic Aryans of the plains pushed upwards, and another branch of Aryans Khasha might have penetrated into the region gradually marching through the hills of Kashmir. In the course of centuries of pressure and upheaval, the Kinners, probably along with Kirats in the adjoining region were gradually pushed towards higher hills, confining themselves only to a very small region now known as Kinnaur district of Himachal Pradesh.[24]

Originally these people may have been nomads. In course of time they came to settle down. After the settlement there must have developed some kind of control over the land. Those who came to exercise this control may have become the rulers, whom we know today as Thakurs. In each small unit, deity, Thakur, and people came to settle. The combination of different ruling people must have wielded political power as they were known to grammarian Penini as *Janapadines*, and their territories as *janapada*. It seems, therefore, that the earliest traditional organized political or social life in these hills was that of *janapada* type, and the area being inaccessible to a greater extent the same type of organization continued up to a much recent period. Perhaps there is much truth in the claims of certain ruling families who trace their origin in a much remote past going back to *Mahabharata* times.

NORTHERN INDIA FROM THE SIXTH CENTURY B.C. TO A.D. 647

It would be worthwhile to look at this region of Kinnaur along with the general conditions of northern India, particularly the hilly regions of Himalaya during the course of several hundred years till the death of Harsha in A.D. 647. After the long traditional epic period which we recorded, we now come on firmer historical ground. In the beginning of the sixth century B.C. India was divided into sixteen great Janpadas and several smaller ones. Among the Gandhara, Kamboja, Kuru, Koshal, Mull, Vajji, Panchal, Sakya, were either in the southern Himalayan regions or had territories extended up to the Himalayan ranges. The extent of the country known to the great grammarian Panini is indicated in the Sutras. According to that information the western-most point of Bharatavarsha was Praknava, corresponding to modern Ferghana. To the south of it was Kamboj the modern Badakshan Pamir. It was known as Gandhara in the contemporary literature and the edicts of Ashoka. The extent of Gandhara was roughly around Rajouri including Hazara district in Pakistan, and probably extending as far as Kafiristan. South of Kamboj was Kapisi, the modern Kafiristan south east of the Hindukush. The Kabul River valley with its outpost at Takhshila was known as Gandhara Valley. At times Kashmir was also included in Gandhara. Among the autonomous clans the Sakya state was the most important. It was bounded on the north by the Himalaya, on the east by the River Rohini and on the west and south by Rapti. Besides these, Trigarta, i.e. Central Kulu, Mandamati, i.e. Mandi, Kalakuta, i.e. Kumaun were also known. Kalakuta was an extensive region including the lofty mountains, wherein the Beas, the Sutlej, the Yamuna and the Ganga had their sources.[25] Perhaps to this Trigarta Kinnaur was a part.

Among the states that were flourishing in the sixth century B.C. the kingdom of the Magadha was the first to make a successful bid for supremacy under Bimbisara and had left behind a rich legacy of imperial tradition. Its emperors belonging to Sunga, Nanda, and Maurya dynasties carried their banners up to the inhabited parts of inner Himalayan regions. Chandragupta Maurya who freed India from the Macedonian yoke brought about its political unification under one sceptre, negotiated an alliance with Parvataka (Himalayan King) before empire building. With the help of several frontier tribes such as the Kiratas, Kambojas, Panasikas and Valhikas, he built up the great Mauryan Empire. The empire of Ashoka extended up to the natural boundaries of India and beyound that in the West. His influence spread far and wide even in the north Himalayan regions. His son Kunal, was the Governor of Gandhara and another son, Jaluka, became an independent ruler of Kashmir after his father's death.[26] After the collapse of the Mauryan Empire the Kushanas established an extensive empire within and beyond India in the north-west. Emperor Kanishska's hegemony spread over Kashmir and the Central Asian regions of Kashgar, Yarkand and Khotan. In Central Asia he fought against China for a long time to maintain the integrity of his frontiers with China. In the Central sector of the modern boundary of India in the north, his hold extended up to the territory of the Inner Himalaya. Kinnaur must have been the part of this empire. In the meantime northern India was divided into a number of small kingdoms and autonomous tribal states. From Kashmir to Kafiristan there was a great *janpada ki-pin*. In the neighbourhood of Jamuna, Sutlej and Beas, the Kuninda tribe was ruling. Kulu was inhabited by Kulutas. Territory to the east of Kangra was occupied by Audumbaras. Nagas were the rulers between the Ganga and Jamuna valleys in the north. Rohilkhand and the adjoining areas were parts of the kingdom of Ahichchatra.[27] Under such a divided country the Gupta Empire grew. Samudragupta's Empire included the territories of Rohilkhand, Kumaun, Garhwal, Nepal, and Assam. Its northern boundary was along the high Himalayas. Kinnaur must have been included in it too. Chandragupta II Vikramaditya (A.D. 375-414), the son and successor of Samudragupta had completed the task begun by his father. His victorious arms penetrated to the far eastern limits of India, and beyond the Hindukush in the North West, The Gupta Empire continued to flourish up to A.D. 467.[28] On the disruption of the Gupta Empire Mihirkula, the Huna leader, established his sway in Kashmir and the neigbouring areas, but was dislodged within a few years.[29] Kinnaur might have been part of this empire. Yasodharman of Malwa had a meteoric career between

A.D. 530 and 540. His suzerainty was acknowledged over the vast area extending up to the Himalayas in the north. The erstwhile rulers of Bashahr may have been concerned, to the ancestors of this Yasodharman.

Early in the seventh century Harsha came to power at Thaneswar in A.D. 606. During the course of the next four decades he had established a most powerful empire in India. All the existing kingdoms of Kapisa, Kashmir, Kuluta, Satadru, Mo-li-pa-lo and Suwarnagotra were incorporated in his empire. Mo-li-pa-lo was the name of Ladakh and Suwarnagotra was situated in the high Himalayas. It was ruled by women and was known as the kingdom of women. The sovereign sway of Harsha extended from the Himalayas to the Vindhyas including Kamrupa. Harsha was probably the second great emperor of India who had diplomatic relations with China.[30] He died in A.D. 647 and with him the glory of Thaneswar and Kanauj departed for ever. The country was once again divided into the old principalities of the sixth century B.C. with the addition of Gurjara-Pratiharas, Guhilots and Chapas. Thus it is evident that till this period there was no Kinnerdesh, nor did Kinnaur have any independent existence of its own.

It appears that the princes being adventurous owing to territorial greed, first went to these high hills and established themselves on varying territories in accordance with their powers. Particularly the area between Sutlej, its tributaries and Baspa up to Mansarover was under the rule of Thakkars from very early times. They were known by the place they had occupied like that of Chini Thakkar and Kamru Thakkar, under ths overall suzerainty of Mauryan and Gupta Kings later on. It was the Thakkur of Kamru who proved strongest of all the other Cheifs of the area and annexed their territories by force sometime after the fall of the Kanauj Empire and laid the foundation of the state of Bashahr, to which the region of Kinnaur belonged till the dissolution of the state very recently.

BASHAHR

It is very difficult to formulate any opinion regarding the foundation of this state on the basis of legends available concerning its origin. According to one legend, Banasur, a contemporary of Shri Krishna was the founder of the state, while another says it was Pradumna, the son of Shri Krishna who founded the state. There is yet another legend according to which the ruling clan (*Rajavansa*) is said to be of Brahmin origin. This suggests that these were the Brahmin *pandits* who must have concocted these stories at different times, to please the Raja Vamsha.[31] In one of the earliest documents available it is said that, 'This Raj was founded by an emigrant Rajput from the Deccan in Sumwat 472, by the name Dunber Singh.'[32] Probably he was one of the ancestors of Yasodharman of Malwa, who rose to power in A.D. 530 which is Sumwat 587.

In Kinnaur each village has a deity. Every deity has a worshipper (*pujari*) and a mouthpiece of the deity. History usually narrated at the time of fairs and festivals through the *grockch* or mouthpiece (a person in a trance.) These utterances are called *chironigns*. The utterances of the Kamru deity along with a few others are given below:

> Generally it says that one Dev Purna went to Kamru and after killing the ruling Thakur of the village and that of the adjoining village of Sangla proceeded to Tangling. The Tangling Thakur was also liquidated after which Dev Purna went further on Chini (near Kalpa) the present district headquarters. The Chini Thakur named Amuruch was liquidated and his fort captured by Dev Purna. Having achieved this, the remaining Thakur at Choling was also done away with by Dev Purna and after subjugating the whole area he proceeded to Sarhan. In Sarhan Banasur was the ruling Chief with whom Dev Purna waged a great battle in which Banasur was defeated and vanquished. Thereafter Dev Purna was alleged to have gone to Kashi Nagar (probably Varanasi of today) and brought from there a man named Pradumna who belonged to Chandravanshi dynasty and installed him as King at Sarhan. What happended to Dev Purna thereafter has not been related in the Chironin.
>
> In another legend Dev Purna is said to have come from Badri Nath in Garhwal through the Himalayas and first halted at Tholing Muth, a place in Tibet. There Dev Purna did not feel at home and is said to have escaped to reach Kamru. His battles with the Thakurs who were occupying different parts of the area have been described in detail and the legend ends at a place where after ridding the whole area from the hated rule of the merciless Thakurs

> Dev Purna set Praduman on the throne at Kamru and himself assumed the role of a deity. Thus the rule of Praduman commenced in Kinnaur with his capital at Kamru which, with the passage of time, grew into the erstwhile state of Bashahr. It is also said that along with Praduman his brother Dusrath was settled as a royal priest.
>
> Yet in another legend it is said that these brothers while roaming about in Sarhan visiting the temple of Kali, happen to enter the palace gate at a certain time: the younger one, that is, Praduman was the first to enter the gate and the throne was at that time lying vacant. His elder brother had to be content with the office of the priest of the royal family.
>
> According to yet another story Praduman came to Sarhan in order to marry the younger daughter of Raja Bavasa Dev (or Banasur). It is not known whether the marriage took place but Praduman is said to have killed his prospective father-in-law and usurped his powers at Kamru.[33]

In all these accounts who was Dev Purna or elder brother? Can we not identify him with Dunber as appears in one of the documents? The Singh in his name may have been added at a later stage as we do not find it with Yasodharman either. It might have come into use later at the time of the Rajput dynasties. It is just possible that the estate was conquered and consolidated by Dunber or Dev Purna and its first important ruler was Praduman. The successors of Dev Purna or Dunber must have continued to rule this region in isolation. We have a very long genealogy of the rulers of Bashahr both at Kamru and Rampur.[34] It would be wrong to say that the region at later stage was ruled by Bhot rulers. For the Kullu and Chamba *Vanshavalies*, it appars that the state had its independent existence at least up to the tenth century, which incidentally is the period of the first Bhot Empire.[35] The second Bhot empire perhaps never existed, and rulers of Ladakh never came below Sutlej. The extent of Ladakhi empire extended only up to Spiti. Therefore, it is necessary to examine the nature and extent of Tibetan imperialism.

TIBETAN IMPERIALISM

In Tibet Sangtsen Gampo was a great king. He came to power in A.D. 618 and soon set about the consolidation of the kingdom. His successor, Mang-Song Mang-Tsen (649-676) soon began his imperial career. He came into conflict with Turks and Chinese power. After subduing them Mang-Song Mang-Tsen extended his empire beyond the Pamirs, including the whole of eastern Turkistan. During the time of his successor, Do-Song Mang-po-Je (676-704), the four garrisons (Kashgar, Khotan, Kuchu, Karushsher) were reconquered by the Chinese in A.D. 692. His entire career was spent in suppressing rebellions in different parts of the empire. He took the field in person against Nepal and was killed during the course of a campaign in A.D. 705. Tri-de-Tsung-sen, successor of Do-Song Mang-po-Je, directed his attention towards China and consequently the border people once again became independent of imperial authority. Being free from China, he once again turned towards the West, and recaptured the four garrisons of Sinkiang from where the Tibetans had been driven out in A.D. 692 by the Chinese forces of the Emperor Wu. The Tibetan army advanced westward to the Pamirs and even reached the Oxus river. A few years later, the Arabian Caliph Harun-Al-Rashid allied himself with the Chinese in order to keep the Tibetans in check. Attacked by the allied forces of the Arabs and the Chinese, the Tibetans succeeded in holding their own without substantial loss of territory despite considerable setbacks. The expansionist dreams, of the Tibetans were, however, checked. Soon, afterwards the Tibetan nobility became rebellious and more powerful during the region of Tri-de-Songtsen (800-815). He, therefore, decided to free the monarchy from the power of the nobility, and confiscated their lands. He also divided the country of Tibet from the country of Ragy and defined the frontiers. His definition of the frontier was: 'All the grass under Tibetan rule grew with points looking towards Tibet'.[36] It is to be noted that upper Kinnaur is an arid zone and no gress of any kind grows there. Imperial Tibet once again returned to religion and the feudal supremacy was established. 'The Tibetan monarchy, although it filled two centuries of the history of Asia did not leave any political or ethical traces in Turkistan; it left only scanty and unimportant traces in Tibet itself, which in 842 found itself nearly in the same conditions as it was in the 6th Century'.[37] As regards Ladakh, it was colonized by the Tibetans, but from there too they had to withdraw early in the ninth century. In the words of Petech:

> Ladakh did not constitute an integral part of the Tibetan state, but must have been considered as a dependency or even as a kind of colony, since, like the whole of Western Tibet, it remained outside the territorial organisation of the Tibetan army as described in the *Padma-bkai-t'anyio*, Part V, Chapter 4. This colonial or semi-colonial status is quite natural, because Ladakh's population was not as yet or was only beginning to become Tibetan—a process that must have required a long time. . . when Skyid-Lde-Ni-Ma-Mgon early in the 10th century founded the Western Tibetan Kingdom, he found no trace of Tibetan rule in Ladakh. The lower part of the valley was divided into a large number of very small states, while upper Ladakh constituted a single state a little more important; its dynasty boasted, as did the Gru-gu dynasty, of descending from Kesar. Probably this situation had existed from very ancient times, notwithstanding the invasions the country had suffered.[38]

It appears that the Tibetan thrust was only northwards along the river Indus leading them to Sinkiang. They never crossed the Sutlej and hence there was no Tibetan rule in Kinnaur. Had they ever crossed Sutlej and proceeded to Kinnaur, what stopped them from reaching the fertile plains of Punjab, and why had they taken the trouble to proceed across the entire mountainous region of Ladakh and Sinkiang? The corroborative evidence of the Vamsavalis of Kullu and Chamba also establishes the existence of an independent Bashahr state till the tenth century, which is considered the first phase of Tibetan rule in Kinnaur. From cultural and religious points of view, even now we do not have the Bhot influence in the interior regions of Kinnaur. It is only on the borders of Tibet with Kinnaur that has the influence of Buddhism and Bhot culture. How is it that after a rule of about eight centuries, the Bhots have not been able to influence Kinnaur culture? To me it appears the assertion of Rahul Sankrityayan which had found place in the *Gazetteer of Kinnaur* regarding the Bhot rule's first phase is thoroughly misleading and erroneous.

The second phase of the Bhot rule beginning from the tenth century in Kinnaur also needs careful scrutiny. Tibet after the collapse of its empire in A.D. 842 settled down within the borders and engaged its attention mostly on internal religious affairs. On the north of Kinnaur the first Ladakhi dynasty was founded by Skyid-Lde. He was known by the name of Ni-Ma-Mgon. He was followed by his son, Lde-dpal-nkor-btsan. The period from 842 to 900 was known for its vigorous religious activities. Skyid-Lde-Ni-Ma-Mgon (900-930), the third in the dynasty was a great king. He consolidated the kingdom of Ladakh and his possessions were far and wide. The kingdom of Ladakh in the time of Skyid-Lde-Ni-Ma-Magon extended from the northern mountain ranges up to Roduk, including Demchog, Guge (South east of Roduk and West of Mansarover) and the modern districts of Lahaul and Spiti. The king had three sons. Before his death he divided the vast kingdom of Ladakh amongst his sons : to the eldest Dpal-gyi-mgon, he gave Ladakh proper; to the second Purang and Guge were assigned; and the third, Lde-Gtsug-mgon, was made the ruler of Zanskar and Spiti. Dpal-gyi-mgon, the owner of Ladakh, was made suzerain over his brothers.[39] After the partition Ladakh enjoyed complete peace and prosperity up to the fourteenth century, except for a minor attack by Kublai Khan in 1207. Utpala (1090-1110) had subjugated Kulu, Purang, Guge and several localities of Baltistan in the eleventh century. Rgyal-Bu-rin-chen (1320-1360) in the fourteenth century was a great king of Ladakh. He invaded Kashmir and overthrew the ruler Sahadeva.[40] Thus the activities of Ladakh rulers never came below Spiti. To trace any kind of Bhot rule in Kinnaur in the second phase also seems to be as much misleading as that in the first phase.

All along history till the fourteenth century Bashahr had its independent existence. The rulers and ruled lived in perfect peace in complete isolation from the events of the plains of India or from the upheavals of beyond Kinnaur's borders with Tibet, Ladakh, and Kashmir. There must have been a loose control of the Raja of Bashahr over the region of Kinnaur, which was called *Sat Khund*. During the several centuries from the establishment of State of Bashahr in A.D. 415, people gradually attained certain degree of prosperity. Border trade was developed and agricultural progressed. Side by side the consolidation of Bashahr state continued. In course of time its capital was shifted from Kamru to Sarhan and later to Rampur. But according to tradition the ruler had to visit Kamru for his second enthronement ceremony invariably.

In the medieval period, though some of the hill states such as Kangra, Chamba, and Sirmaur were attacked and made tributary to the Mughal Emperor at Delhi, Bashahr state could not be reached by any advanturer of that period. The consolidation and addition of territories to the Bashahr state continued during that period also. The Thakoorais of Dulaitoo, Kurungoloo and Kuaitro were annexed about Sumwat 1611.[41] Perhaps he was Raja Chatar Singh, who brought the whole area of the erstwhile Bashahr state under his control. He was considered as a most virtuous ruler during his reign. Nothing particular is known about his successor Kalyan Singh.

The successor of Kalyan Singh according to genealogy was Raja Kehri Singh. He is described as the highest skilled warrior of the time. He was easily the most powerful chief in Western Himalaya in his times. He is said to have reduced the Rajas of Sirmaur, Garhwal, Mandi and Suket to his vassals, and to have curbed the Thakurs of Keonthal, Kotkhal, Kummarsain, Balsan, Theog, Dinkoti, etc., and wrested Rawn from Garhwal.[42] During his time the Mughal Emperor Akbar held a great darbar of the hill chiefs, which he also attended. It is said that it was a hot and bright day, and wherever Kehri Singh went a small cloud formed a shadow for him. People who saw this were surprised and subsequently Akbar came to know about it. Akbar asked the Raja the reason for this phenomenon. 'This is on account of the blessings of the hill gods' was the simple reply of the Raja. The emperor was very pleased to hear it. He gave the Raja a Khilat and the title of Chhatrapati saying that since the Raja was given a *chattar* (umbrella) by the gods he described the title. Kehri Singh continued to be known as Chattrapati since then. In the Tibetan-Ladakhi-Moghal war of 1681-1683, Raja Kehri Singh sided with Tibet. The importance of his role in this war was very crucial. At the end of the war Tibet rewarded him for his help by giving him all of upper Kinnaur beyond

Sutlej which was occupied by them during the war. It also signed a commercial treaty conferring on traders from Bashahr the right of free trade and movement in Tibet.[43]

Kehri Singh's successors were not of the same mettle. Besides mention in genealogy of Bashahr state, nothing is known about Vijay Singh and Udai Singh. It is said that one Raja Ram Singh made Rampur his capital, in place of Sarhan and Kamru. During his region a series of contests began with the Raja of Kulu and Bashahr had lost the territory of Seraj. It seems that the territories which were annexed by Raja Kehri Singh became free during the week rule of Raja Rudra Singh. But his successor Ugar Singh took them over by force of arms. He gave them to the neighbouring small states but maintained suzerainty over them. There had been a long standing hostility with Sirmaur State and Bashahr. It was only after the marriage of Ugar Singh with the daughter of the above state that the old dispute came to an end. The Raja on his death left two Ranis, one from Sirmaur and the other from Dhami. By the Dhami Rani, Ugar Singh had a daughter, and a son who later succeeded him under the name of Raja Mahendra Singh.[44]

While the petty wars among the hill Chiefs were going on there arose three formidable powers in this Himalayan region. The Sikhs became powerful under the celebrated Ranjit Singh. Nepal became a great power under the able leadership of Prithvi Narain Shah. The British East India Company under the able control of Richard Colley Wellesley became the champion of the 'Subsidiary Alliance.'

GURKHA OCCUPATION

The Gurkhas of Nepal had extended their dominions greatly during the end of the eighteenth century and the beginning of the nineteenth. Amar Singh Thapa, the Gurkha leader, went up to Kangra Valley. He was driven from the valley by the superior forces of Ranjit Singh, and those of the celebrated Raja Sansar Chand of Kangra. After being driven from Kangra, Amar Singh established himself at Arki a place in the small state of Bhagal, the chief of which he had driven into exile. Amar Singh then requested Sir Daniel Ochterlony, the British Agent, for assistance in his designs against Kot Kangra, which was promptly refused by the Agent.[45] Sansar Chand, along with the Nalagrah chief, attacked Bilaspur state in 1803. The Raja of Bilaspur invited the assistance of the Gurkhas in repelling the encroachment. The assistance asked for was given by Gurkhas, and their intervention in this quarrel marked the beginning of their conquest of the Simla Hills.

> In Sambat 1868 (A.D. 1810) died Raja Oogar Sain. He left an infant heir, the present Raja, 18 months old, and a few months subsequent to his decease the Nepal troops took possession of this state, with the exception of Kunawar, to which place the infant Raja with his mother and the hereditary Wuzeers fled. The Nepal troops followed them as far as Wangtoo, where there was a wooden bridge, called a Sanga, across Sutlej. A partial action occurred between the Kunawarees and Goorkhas, which was followed by the retreat of the latter, but not until the bridge had been destroyed by the inhabitants of the country to cut off the communication with Bushahir proper, which remained in the hands of the Goorkhas.
>
> The Bushahr Prince with his mother remained in Kunawar, but they paid Rs. 12,000 tribute per annum, for being left in peace to the Goorkhas, until the advance of the British troops, when they returned to the Capital in Bushahr.[46]

While the Raja was on his flight, the Gurkhas looted the treasuries and completely destroyed the records of the state. There is also a story that the then Paw Wazir Fateh Ram handed over to the Gurkhas several strong boxes securely locked and filled with stones, claiming that they contained the state treasure. The Wazir told the Gurkhas that he had not been able to get the keys. The Gurkhas readily believed him and carried away the boxes without breaking them open. They did not discover the trick played upon them until too late.

The tract between the Sutlej and Jamuna came under British protection by the Treaty of 1809 between Ranjit Singh and the British Government. Thus the British Government took positive steps to expel the Gurkhas and on 1 November 1814 war was declared. Colonel Ochterlony, with the support of the local chiefs, marched into the hills. After a long and desperate struggle Ochterlony completely defeated Amar Singh Thapa on 15 April 1815 and a convention was signed on

13 May 1815 which was later confirmed by the treaty of Sangauli on 2 December 1815.[47]

On the conclusion of the Gurkha War, Raja Mahendra Singh was granted a *sanad* on 6 November 1815. It gave Khaneti and Delath Thakurais to Bashahr, and a part of Rawain, which was a district of the state, was transferred to Keonthal; Kumharsain was constituted a separate Thakurai. The Chief of Bashahr was to afford subordinate military co-operation whenever called upon, and to furnish *begaris* for the construction of roads throughout his territory, besides a tribute of Rs. 15,000.[48] This tribute was however reduced to Rs. 3,945 as a compensation for the abolition of transit duties in 1847.

Next in importance after the Raja family were the three Kanit hereditary Wazir families, named *Pawari Konal*, and *Shuna*. The Pawaris being the oldest family were descended from an able head called Nargu, who came from Kullu in attendance on a Rani belonging to that region and was made Wazir by the Raja. In this family the best known names are Manohar Das and Ran Bahadur Singh. The former appears to have been a man of some ability, who was all powerful previous to the administration of Tikka Raghunath Singh. The *Konals* came from Garhwal some seven generations ago. The *Shuna* Wazirs were natives of the state. The members of these families rose to more importance during the time of Raja Mahendra Singh who was not a capable man. The influence of these hereditary Wazirs was paramount in the state until the administration was entrusted to Tikka Raghunath Singh. Previous to the Gurkha War the Rajas had kept a firm personal grip upon the state affairs, and the Wazirs had occupied their legitimate positions as counsellors. There was one Wazir who was invested with considerable independent authority but his post was not hereditary. He was *Sarhaddi* Wazir, who was in charge of the frontier. He used to be selected amongst the able men of the parganas of Shua or Tukpa, and was practically given a free hand as long as he maintained order. No member of the above mentioned three families ever held the frontier Wizarat. This post was not revived after the expulsion of the Gurkhas.[49]

Raja Mahendra Singh was a minor when the *sanad* was granted, and on coming of age he proved a very weak ruler, so the state was administered by the Rani, his mother, and the hereditary Wazirs under the direct supervision of Sabathu Agency. In 1850 Shamsher Singh, a boy of eleven succeeded to the throne (*gaddi*), and during his minority the state was first administered by Mansukh Das Wazir as Regent, and later by Shamlal, Tahsildar of Nurpur, as manager under the direct control of the Agent for the Hill States. During the 1857 insurrection, the Raja kept back his tribute and offered no aid to the British Government. He treated officials travelling through his territory with discourtesy and refused even ordinary supplies to them. Lord William Hay, the Deputy Commissioner Simla and Agent for the Hill States, proposed to send a force to Rampur to coerce him but there were no troops available. He also proposed that the Raja be deposed and the state should be taken under the direct management of the Superintendent, Simla Hill States. This was not however, deemed advisable by Sir John Lawrence, the then Chief Commissioner of the Punjab, and the Raja's behaviour during the Great Revolt was practically ignored. Except that the Superintendent, Simla Hill States, repeatedly intervened in the internal feuds of the state to prevent them from affecting the smooth running of the administration. In a typical dispute between Bashahr and Mandi over the marriage of Tikka Raghunath Singh of Bashahr to the two daughters of the Raja of Mandi in 1883, the Superintendent seized Wazir Ram Bahadur Singh, who went to Sirmaur with Tikka Raghunath Singh, and removed him to Simla. The crisis was due largely to Raja Shamsher Singh who was a weak-minded person. Tikka Raghunath Singh finally accepted the marriage alliance with Mandi.[50] During this period the administration of the state was in a very bad shape in the hands of hereditary wazirs.[51] Raja Shamsher Singh was induced by the Superintendent to delegate his full powers to his only legitimate son Tikka Raghunath Singh in 1886,[52] whose administration was a complete success till his death on 2 February 1898.[53] During this period the power of the hereditary wazirs was reduced. The first land settlement was done by Durga Singh under the direct supervision of Tikka Raghunath Singh in 1893-94.

Soon after the death of Tikka his eleven-month son also died in 1899. The direct line was feared to become extinct, as Raja Shamsher Singh was seventy years old. He adopted Surendra Shah, brother of the Raja of Garhwal as his son and successor to the *gaddi*. The only survivor, a son by a concubine, Padam Singh was illegitimate. Several claimants advanced their claims to the throne (*gaddi*), particularlay the daughter of Tikka Raghunath Singh who was married to the Raja of Kashipur and had a minor son. The British Government first recognized the adoption but subsequently cancelled it due to the complicity of Surendra Shah in an attempted murder of a forest officer Mr. Gibson. Due to this he was expelled from the state.[54]

Soon after the death of Tikka Raghunath Singh the old Raja Shamsher Singh made an attempt to regain the administration, but the British Government intervened and appointed Rai Sahib Mangat Ram as Chief Wazir with the same powers as had been wielded by Tikka on the salary of Rs. 100 p.m.[55] The Raja had then practically nothing to do with the administration of the state. In 1899-1900 the management of Bashahr continued satisfactorily under the new manager, and by the exercise of rigid economy the debts of the state were reduced to just Rs. 50,000. The Raja was incompetent and occasionally obstructive. He neglected to meet the Superintendent Simla Hill States at Nogri Nullah on his visit to the state in the autumn of 1898-99 and was fined Rs. 1,000 as a punishment for this mark of disrespect to the representative of the paramount power.[56]

Raja Shamsher Singh died on 4 August 1914 and was succeeded by Padam Singh on the next day. Padam Singh had been trained in administrative matters for three years under the supervision of one Mr. H.W. Emerson, I.C.S., who had served as manager of the state during 1914-17. The cost of his establishment was Rs. 18,000 a year. Padam Singh was given full powers of state on 30 November 1917.[57] The Raja of Bilaspur applied the *Raj Tilak* on the forehead of Padam Singh. He was recognized by the fraternity of Rajput chiefs in the Hills. The ceremony was attended by several hill chiefs. Bashahr at last came under an administration which was certainly progressive in comparison with that of the other neighbouring hill states. Between 1939 and 1946 many administrative reforms were introduced, with the emphasis on the efficiency of the judiciary, police, education and public health. Padam Singh was married in the house of Dhandih and Lambagaon (Kangra District). He was still holding the reins of Government when India became independent in 1947. Raja Padam Singh left behind two sons, Raja Vir Bhadra Singh and Raj Kumar Devinder Singh. Vir Bhadra Singh was a Member of Parliament and a Deputy Minister in Mrs. Indira Gandhi's Government till quite recently.

FOREST LEASE

Forests were the greatest source of revenue for the state of Bashahr. Although a scheme for taking over the forests was prepared in 1850 by the British Government, it could be effected only in 1864 when the Raja of Bashahr leased his forests, for a period of fifty years. The final single agreement of 1877, besides securing to the Raja an annual payment of Rs. 10,000 in consideration of the rights conceded, made more definite provision for the conservancy of the forests leased. In 1905, keeping in view the bad finances of the state it was decided to pay off the debts amounting to Rs. 27,000 and a provision was also made to give Rs. 30,000 annually for ten years for road construction and general administration out of the profits of the forest of Bashahr.[58] In 1929 a fresh agreement in regard to forests was made. This was to continue for twenty five years. Under this certainly definite provisions for the conservancy of the forests were made, and the annual payment to the Raja was raised to Rs. 1,00,000. It is to be noted here that between 1864 and 1904, after expenditure, the British Government gained Rs. 7,18,843.[59]

POPULAR REVOLTS

Popular revolts against the tyranny of the state were not unknown in the state of Bashahr. A revolt headed by Fateh Singh began in 1859 when the people left the fields and took shelter in the hills. This was against the cash payment of land revenue, which was introduced by Munshi Shamlal in 1854 and the removal of Paras Ram, a Vakil, who had been nominated as Wazir of the State. G. Barnes, Superintendent of the Hill States, was obliged to proceed to Bashahr to settle matters, and as a result of this the old system of payment of revenue in kind was restored. Another revolt headed by Ran Bahadur Singh, one of the hereditary Wazirs, continued from 1899 to 1906. He was trying for quite sometime to establish Dodra Kawar in Rohru Tahsil as an independent principality under himself. In the summer of 1906 the condition became worse. A police force under the Political Assistant Commissioner, Simla was despatched and Ran Bahadur was arrested and lodged in Simla jail for two months, and was released only when he agreed to pay land revenue.[60] Ran Bahadur Singh has been mentioned as a nationalist of those early days and was possibly known even outside the state of Bashahr in his times.

After the tragedy of Jallianwala Bagh on 13 April 1919, and under the influence of the Non-cooperation movement an agitation was conducted in the Simla Hill states against the ancient custom of *begar* (forced labour). The Punjab Government undertook in the winter of 1920-21 a complete revision of the existing system on the Hindustan Tibet road between Simla and Bashahr. The use of *begar* was practically abolished. The Prajamandal movement for the freedom and a responsible government was continued. After Independence the Simla Hill states rulers made a last and unsuccessful attempt to survive as independent entities in early February 1949. The chiefs met at Solan under the chairmanship of Raja Durga Singh of Solan (Bhagat) to draft a constitution to form a separate union of states. But the attempt failed for the want of popular support. To this effort of the Rajas, the Prajamandal retaliated with seven days *Satyagrah* which shook the states of Himalaya.[61] The provisional government of Himalayan Prant was formed with Shivanand of Sirmur as its President and Sadanand Chandel of Bilaspur, Pandit Padam Dev of Bashahr, Mukand Lal of Suket as its members. It remained only till Sardar Vallabhbhai Patel merged thirty one princely states to from a part 'C' State under a Chief Commissioner on 15 April 1948. On 24 March 1952 first popular ministry headed by Dr. Y.S. Parmar was established. In 1954 Bilaspur was also merged with Himachal Pradesh. In the meantime the States Reorganisation Commission was appointed on 29 December 1953 with Dr. H.N. Kunzru, S. Fazal

Ali and Sardar K.M. Panikkar as its members. Under its recommendations Himachal Pradesh was finally established in 1955. On 1 November 1956 the Part C state of Himachal Pradesh was made a Union Territory under a Chief Commissioner. The popular ministry was installed under Dr. Y.S. Parmar on 1 July 1963. The remaining hilly regions of Punjab were merged with it on 1 November 1966 and finally it became a full-fledged state on 25 January 1971. Kinner region of Bashahr state was first merged with Mahasu district but on 21 April 1960 the Chini Tahsil of Mahasu district was made the new district of Kinnaur. Thus the mythological Kinnerdesh became Kinnaur district in modern times.

REFERENCES

1. Kalhana, *Rajtarangini*, ed., Raghunath Singh (Varanasi, 1969), p. 110; see also Bhagwat Saran Upadhyaya, ed., *Hindi Viswakosh* (Varanasi, 1963), Vol. 3, p. 6.
2. D.P. Misra, *Studies in the Proto-History of India* (Delhi, 1971), p. 20.
3. R.K. Mukherji, *The Original Inhabitants of India*, p. 514.
4. *Matsya Purans*, VI-45.
5. F.E. Pargiller, *Ancient Historical Traditions* (Delhi, 1962), p. 186.
6. *Vishnu Purana*, 2:1-11-19, (Gorakhpur, 1953), p. 136.
7. S.N. Vyas, *India in the Ramayana Age* (Delhi, 1967), p. 186; Alain Danielo, *Hindu Polytheism* (London, 1964), p. 26, 306.
8. A.C. Burnell, *The Ordinance of Manu* (Delhi, 1971), p. 68.
9. Vyas, n. 7, p. 25; Mishra, n. 2, p. 20; *Matsya Purana*, n. 4, CLIII, pp.28-37; B.S. Upadhyaya, *India in Kalidasa* (Delhi, 1963), pp. 2, 7, 62, 298.
10. *Vishnu Purana*, 2:2-47 and 2:4-37.
11. *Shiva Purana*, 21-36-38 (Delhi, 1970), Vol. 1, p. 145.
12. *Mahabharat*, *Vana Purva*, 27-33, 53-9, 105-21; *Ashvamadhik*, 88-379.
13. *Adi Purana*, 19-33.
14. P.V. Bapat and A. Hirakuwa, *Shan-Chien-Pi-Po-Sha : A Chinese Version by Sanghabadra of Samantrapasadika* (Poona, 1970), p. 9. p. 215.
15. E.B. Cowell, ed., *The Jataka or Stories of the Buddha's Former Birth* (London, 1969). Also see Heinrich Aimmer, *Myths and Legends* (New York, 1953).
16. Ananda Coomaraswamy, *Yaksas* (Delhi 1971), pp. 6, 10.
17. R.C Agrawala, "Yaksi or Kinnaris Pot from Begramin", in Amlendu Guha, ed., *Central Asia* (Delhi, 1970), pp. 133, 172.
18. Rahul Sankrityayan, *Himalaya Parichaya* (Allahabad, 1953), p. 42.

19. Chatursen, *Bhartiya Sanskritika Itihas* (Meerut, 1958), p. 35.
20. Budh Prakash, *Studies in Indian History and Civilization* (Agra, 1962), p. 356; and S.K. Chatterjee, *Cultural Heritage of India* (Calcutta, 1958), p. 41.
21. S.M. Ali, *Geography of the Purana* (Delhi, 1966) p. 53.
22. P. Thomas, *Epics, Myths and Legends of India* (Bombay, n. d.), p. 19.
23. *Raghuvansa*, IV-77-78. Up to this portion, the material is mostly drawn from an unpublished paper written by Mr. Govardhan Singh, Librarian, Secretariat Library, Simla, Himachal Pradesh.
24. Rahul, n. 18, p. 42.
25. S.C. Bajpai, *The Northern Frontier of India* (Bombay, 1970), pp. 4-5. See also V.S. Agrawala, *India as known to Panini* (Lucknow, 1953) pp. 37-54; R.C. Majumdar and A.D. Pusalkar, eds, *The History and the Culture of the Indian People : The Age of the Imperial Unity* (Bombay, 1951), Vol. II, pp. 1-17.
26. Ibid., p. 5. See also Romilla Thapar, *Asoka and the Decline of the Mauryas* (Oxford, 1953).
27. Ibid., pp. 5-6. See also Majumdar and Pusalkar, eds, n. 25, pp. 122-61.
28. Ibid., p. 6. See also Radha Kumud Mookerji, *The Gupta Empire* (Bombay, 1948),
29. Ibid., p. 6. See also R.C. Majumdar and A.D. Pusalkar, eds, *The History and the Culture of the Indian People: The Classical Age* (Bombay, 1954), pp. 34-95.
30. Ibid., p. 6. See also R.S. Tripathi, *History of Kanauj* (Banaras, 1937); Majumdar and Pusalkar, eds. n 29, pp. 96-123. One can also see, L. Petech, *Northern India According to Shai-Chung-Chu* (Rome, 1950).
31. R.K. Kaushal, *Himachal Pradesh : A Survey of History of the Land and its People* (Bombay, 1965), p. 46.
32. From Captain O.P. Kennedy, Assistant Deputy Superintendent, Sikh and Hill States, to Lieutenant W. Murray, Deputy Superintendent, Sikh and Hill States, dated Sabathu, 6 July 1824, Punjab Government Records, *Delhi Residency and Agency, 1807-1857* (Lahore, 1911), Vol. I, pp. 255-99. Hereafter to be called Kennedy Report, 1824.
33. M.D. Mamgain, ed., *Gazetteer of Kinnaur* (Ambala, 1971), pp. 53-4
34. Ibid., pp. 366-68; Rahul Sankrityayan, *Kinner Desh* (Allahabad, 1956), pp. 245-48.
35. Kaushal, n. 31, pp. 46-47.
36. Bajpai, n. 25, pp. 10-11. See also A.H. Francke, *Antiquities of Indian Tibet* (Calcutta, 1926), Vol. II, pp. 37, 76-8.
37. Luciano Petech, *A Study of the Chronicles of Ladakh* (Calcutta, 1939), pp. 55-85; Francke, n. 36, pp. 84-9; Tsepou W.D. Shakabpa, *Tibet : A Political History* (London, 1967), pp. 23-53.

38. Petech, n. 37, pp. 102-3; see also Bajpai, n. 25, p. 11.
39. Petech, n. 37, p. 108; Francke, n. 36, p. 94; Zahiruddin Ahmed, "Tibet and Ladakh: A History", *St. Antony's Papers*, No. 14 (London, 1963), p. 39. See also Bajpai, n. 25, p. 11.
40. Bajpai, n. 25, pp. 11-12. See also R.C. Majumdar and A.D. Pusalkar, eds, *The History and Culture of the Indian People: Delhi Sultanate* (Bombay, 1960), Vol. VI, pp. 376, 429.
41. *Kennedy Report 1824.*
42. Ibid., para 90.
43. For details see Bajpai, n. 25, pp. 13-14.
44. *Kennedy Report*, para 40.
45. Mian Bashir Ahmed Farooki, *British Relations with Cis-Sutlej States, 1809-1823*, Punjab Government Record Office Publication, Monograph No. 19 (Lahore, 1941), p. 43.
46. *Kennedy Report*, 1824.
47. National Archives of India, *Foreign Department Misc. Series*, Vol. 365, Report on the Administration of India, Punjab and Territories, p. 123. All the documents mentioned are from N.A.I., New Delhi.
48. Foreign Department, Secret, 16 December 1815, No. 12-13.
49. Mamgain, ed., n. 33, p. 61.
50. Col. C. Beadon, Supdt. Hill States to Agent, 6 May 1884, *Foreign Internal A*, January 1886, No. 71.
51. Ibid., January 1886, No. 84.
52. Ibid., December 1886, No. 277.
53. L.W. Dane to Secretary, Govt. of India, 11 February 1898, *Foreign Internal A*, August 1898, No. 133.
54. Memorial to Secretary of State for India from Surendra Shah. *Foreign Internal A*, November 1911, Nos. 81-2.
55. L.W. Dane to Secretary, Government of India, 10 June 1898. *Foreign Internal A*, August 1898, No. 115.
56. Superintendent Hill State to Chief Secretary Punjab, 30 October 1899, *Foreign Internal A*, January 1900, No. 68.
57. K.W. *Foreign Secret Internal*, October 1919, No. 31-4.
58. From A.H. Diack, Chief Secretary, Punjab to Secretary, Revenue and Agriculture, Govt. of India, 22 December 1905. *Foreign Internal B.* September 1906, No. 120.
59. *Punjab State Gazetteer*, Vol. 8, Simla Hill States (Lahore, 1910), pp. 1, 59.
60. *Report on the Administration of the Punjab and its Dependencies for 1906-7* (Lahore, 1908), p. 5.
61. Dr. Y.S. Parmar, 'Seven Days that Shook the Himalayas', *Tribune*, 7 March 1948.

4

Economic Resources

THE MYTHOLOGICAL legendary account of Kinnerdesh does not tell us anything about the source of livelihood of the Kinnaura people. Originally they must have been pastoral. The ecology must have played a great part in the evolution of their professions. In a region,

> surrounded by rocks of difficult access, inimical to industry, offering little inducement to cultivation, in a climate of protracted rigors, the inhabitants of Knnawr have followed that course which nature dictates, and we find them active, enterprising and industrious, occupied in extensive commercial intercourse and trade, rearing vast flocks which form their chief dependence and trafficking into remote countries under great hardships and privation to gain a comfortable subsistence for their families at home.[1]

Therefore it appears that there has never been a single primary source of livelihood of the people of this region. All the avenues of animal husbandry, horticulture, agriculture, forestry, cottage industry, trade and commerce must have been equally important for them, from the earliest known times. This fact was noted by an earliest British traveller James Baillie Fraser in 1814. The conditions did not change even during the visit of Captain Alexander Gerard in 1841. In course of time Kinnaur had developed a closed economy. Since the circulation of money was limited it gave rise to barter economy. Monetized markets were altogether non-existent. With the passage of time some periodic fairs dedicated primarily to trade and commercial activities led to the introduction of a money economy in a most limited sense. During the settlement of 1928 or even at the time of the visit of Rahul Sanskrityayan in 1948 the conditions had almost remained the same except that money lending among the traders had become known and the number of sheep and goats had increased.

ANIMAL HUSBANDRY

Animal husbandry had been and still is one of the main pillars of tribal economy and a profession of great importance. Livestock, especially sheep and goats have played a dominant role here. Besides yielding wool, goats hair, milk, manure and meat, these sheep and goats were used as beasts of burden for carrying goods across high passes often snow . covered and to the plains of Punjab in earlier times. Their importance was much more than bullocks, asses and even ponies.

Chief among the domesticated animals are, horse, mule, donkey, yak, cow, ox, sheep, goat, mare, dog and cat.

Sheep and goats constitute more than half of the livestock. Besides agriculture the major occupation of inhabitants has been sheep and goat rearing. It is still the most important occupation and the mainstay of the age old tribal economy. Manifold benefits are derived from the sheep in the shape of milk, meat, skin and above all wool which besides satisfying the local needs of woollen-cloth is also exported. Similarly goats provide mutton, milk, hair and skins. But they are now regarded as most destructive animals from the standpoint of soil conservation. Therefore, this poor creature seems to be deemed to reduction in numbers if not to total extermination in the near future. In fact a decision was taken by the government sometime back to replace goats by sheep but due to the sharp reaction of the local population, it was decided to go slow in the matter. Here sheep and goats have been used as pack-animals from very early times. Their use as pack-animals is now restricted to only those paths which are considered dangerous for larger animals. In due course of time, it appears that their use as beast of burden will discontinue when the modern transport system will be fully developed in these high regions. Besides the common goat, a special variety called *Chigu* is also found here. This variety is being improved and multiplied due to its valuable fleece called *Pasham*.

Horses, ponies, mules and donkeys are used as beasts of burden. Due to the mountainous tract and non-availability of easy means of communications, the rearing of these animals had been very important. Still their importance has not diminished significantly. Donkey tops in numerical strength followed by horses, ponies and mules.

Yaks are the most important among cattle in Kinnaur. In the higher regions cows are rare and yaks are the substitutes. As a source of milk, the yield of yaks is much more than that of cows. The male is called *yak* or *yag* and the female *breeme*. The yak and the cow are cross-bred and the males of this cross-breed are called *Churu* or *Zo* or *Zofa*, and the female *Zomo*. They are strong and hardy and like the cold climate. The crossbreed is more docile. These beasts are also used for the purpose of transport.

Fishing was known to people from the earliest times, but there were no rules and regulations regarding its practice. During the rule of the Raja a nominal fee was charged from the people for fishing. Besides Baspa and Spiti rivers, certain streams like Barnng, Tangling, Kozang or Kashang, Lippa or Kirang, Shyaso, Leo, Hango, Chuling and Tirung are important from the fisheries point of view. Pond fishing was very rare. With the establishment of a Fisheries Department in 1950-51, greater attention has been given to it. Besides regulations and rules for fishing several measures have been taken towards the development of fisheries. The construction of a new Trout Farm at Sangla was undertaken in 1960-61, and now it is yielding good results.

Poultry farming was unknown to the people of this area. In fact eating chicken was taboo. Now in the changed circumstances the old prejudices are on their way out. Two extension units at Tapri and Peo are functioning at present. The products of these units are used for both table and rearing in the area. Chickens are also available at these centres.

There are smaller land holdings in the region. Fodder was never cultivated in this area. Natural grass has always been a source of fodder in the region. When there is dearth of fodder, cattle are fed by leaves from trees. Fodder available from all sources is not sufficient for the livestock. Therefore, the herds of cattle and the flocks of sheep and goats are let loose to graze in the extensive pastures and during summer months they are sent up to the Alpine pastures. With the start of winter the herds and flocks are brought down to the villages and subsequently migrated to lower parts of the hills and even to the plains of Punjab and Uttar Pradesh. They return to their homes when the days start getting warmer. Stall feeding is not in vogue here, but those cattle left behind during winter are stall fed by hay. The stock of hay is usually supplemented by the stalks and chaff from food crops or leaves from trees on private lands and forest trees. The cattle when in their sheds are also kept like the flocks of sheep and goats, i.e. they are not tethered.

This regular migration of livestock does not indicate that the people were living a nomadic life. It was a settled population, but for those members of the family who were in charge of livestock, particularly sheep, goats, asses and ponies, the existence was not better than a nomadic one. They divided their year between the summer grass lands and the winter pastures and between trading Tibet and grain supplying areas of the plains.

In 1972, when human population was 42,835, the number of livestock against each thousand human beings was 2,552, out of which 1,297 were sheep, 669 goats and 444 cattle, and the gross livestock population was 1,25,708.[2]

IMPROVEMENT MEASURES TAKEN BY HIMACHAL GOVERNMENT

After independence, particularly after the creation of the border district of Kinnaur, in 1960, the state government has taken several measures for the improvement of livestock population. Better breeding and greater output from the livestock are the twofold objectives of the government. A Sheep Breeding Farm at Karchham was started with imported merino ewes from Spain and Soviet Union. Cross breeding and pure sheep breeding are done at this farm. Several rams are kept for breeding during the season and are taken back after the season. It is said that both quality and quantity is much better than the indigenous one. In the beginning five sheep and wool extension centres were opened at Jangi, Saryla, Peo, Kausthal, and Bhabha (Katagaon), but now they have been closed. A *Chigu* goat breeding farm at Sangla was opened in 1960, but was closed later as Chigu goat which was to come from Tibet was not readily available.

A horse and mule breeding scheme has been taken up and a temporary farm is located at Jeori in Simla district. Connemare mares and stallions were imported from Ireland. Spiti mares and stallions as well as Chamurti mares and Tango ponies are also maintained at this farm. For the purpose of mule breeding Jack Stallions have also been purchased. The breeding program is broadly based on the crossing of Spiti and Chamurti mare with Connemare stallions. The Connemara mares are also being crossed with the Connemara stallions.[3]

Poultry farming has also been taken up with the opening of the farms at Tapri and Peo. Poultry farming is a new venture, there never was the practice of chicken eating in Kinnaur. Hence the beginning is most modest.

The Fisheries Department has made efforts to improve the fishing in the district. Now fishing has been brought under a national licensing system. Pond fish culture is engaging a greater attention. Construction of a new Trout Farm at Sangla was undertaken in 1960-61. The object was to raise sufficient number of fishes every year to provide game to the anglers and tourists and for food to the local people.

Animal diseases are now properly diagnosed and treated. To provide necessary veterinary aid to the livestock of the district, the state government has opened nine hospitals at Kalpa, Jangi, Nachar, Puh, Leo, Ribba, Bhabha, Sangla and Kilba. There are twelve dispensaries attached to these hospitals. They are located at Giabong, Chango, Pawari, Chhota Kamba, Rakcham, Nignlsari, Rupi Morang, Lippa, Peo, Barua and Sholdang. In order to protect sheep from skin diseases, sheep dip centres are opened at almost all the routes where the frequency of the sheep is high.

Cattle fairs are now held to market the livestock every year. To inculcate a sense of competition, cattle shows and calf rallies are also organized now periodically.

AGRICULTURE

Though agriculture is the largest source of livelihood of the people, 'The spaces of arable land are few, and the cultivation is commonly in narrow strips along the brows of the mountains. The crops for the most part are poor, and a great want of grain pervades the whole country. . . .'[4] Thus Kinnaur has always been deficient in food grains. That is why such supplementary dietetic items as dried apricot and herbal roots collected from wild growth had to be pressed into service till very recent times. Dried apricot is still in use in many parts. Dried apricots and other herbal roots are grounded in flour before the *chapatis* are prepared. There is no uniform cultivation in the district. It is carried primarily on the banks of the river. The elevation of the cultivable land varies from 900 metres to 1,200 metres at the River Sutlej to 3,650 metres on mountain slopes. The greatest part of the area is beyond the range of monsoon rains, which presents special features of its own. The soil is usually steep and consists of sand, clay, stony and gravelly type. Most of these types are met in each tahsil. Hangrang Valley contains several other types of patches of soil besides sandy loam, stony, and gravelly types of soils. The soils of Nachar sub-division are mostly sandy and stony and is suitable for growing crops like wheat and barley and fruit trees. Sandy loam soil is suitable for lesser millets and for vineyards. As a matter of fact the stony gravelly and sandy soils are not suitable for any crop cultivation; these can only be put under fruit cultivation profitably. But due to shortage of land, there is no alternative but to use it for crops cultivation. Along with this rough soil, its erosion poses a more serious and greater problem. Soil conservation measures on a large scale are the first agricultural requirement. Besides the steep nature of the terrain, water, wind, snow and avalanches constitute the main eroding force and scale of the loose surface soil.

Tikka Raghunath Singh's settlement of 1897 has classified land into several types, such as, *Kiar*, *Bakhal*, *Karali*, *Newal* and *Kandas*. Kiar is first class land. It sometimes yields two crops a year. During *Rabi* rice is cultivated and during *Kharif* maize is grown. This land is irrigated from hill streams. This type of land is mostly available by the riversides, and hence its area is insignificant. *Bakhal* is a second type of land and is available in the surroundings of the village (Guand land as known in plains). This land is manured and ordinarily yields two crops. It is irrigated by (*Khul* system) water carried by a channel to the field, but as a rule it is dependent upon rain. Thus, this type of land is available only in lower Kinnaur. A third type of land is known as *Karali* and it is mostly (*Nautor*) newly reclaimed land. This type of land is at a distance from the village. It can neither be manured nor irrigated. It depends entirely on rainfall. As a rule it yields only one crop. *Newal* land contains vineyards and apricot orchards, and is considered as superior land. It is mostly available on the banks of River Sutlej. It is warm low lying land. It also produces two or more crops a year. *Kandas* are located at a distance from the dwellings and are considered very productive. Above Kapla or more than half of the district is practically outside the purview of rain during summer. Irrigation is most valuable here as elsewhere in the country.[5] The *Khul* system is peculiar to this region. This type of irrigation is essential as there is almost always scarcity of rain. The peculiar topography had compelled people to devise this method. In this, water from natural resources like spring (*dhara*) is carried through channels to the desired place. These channels sometimes are made of wood. Now tin or pipes are also used. Almost all cultivated lands are irrigated through this system now. In areas where even *Khul* water is not available, such crops are raised, as can do without constant supply of water. At places snow water is also utilized for irrigation. When snow starts melting, water is collected and diverted through small check dams and taken to the fields. In the days of the Raja the system was not so effective as it was devised for certain sections of population alone. Now this is based on mass utility. Beyond Lippa aqueducts employed reflect skillful construction. The present government have provided several *pucca Khuls* with pipe lines. This had reduced the expenditure on repairs of Khuls and provided regular water supply.

CROPS

Altitude determines the timing of sowing and reaping of the

different crops. Rabi crops are usually sown from the middle of September to the middle of December and are reaped in April in the lower region. In the higher regions it is July. Kharif crops are sown from March to the middle of July, and reaped from September to the end of November. In the colder parts snow remains on the ground for a long time and, therefore, in some places only one harvest is possible. In some places spring and autumn crops are sown and reaped together. Sowing is done in September and October and harvesting is done in August and September in some areas.

Rabi crops include, barley, wheat, lentils, Coriander seed and field peas. Barley (*tag*, *cha*, *tung*, *nan*, *tak-swa*) is very well adopted to all the climatic regions. It is cultivated on all kinds of soils. It occupies important place as a bread cereal or for processing into a kind of drink. Wheat (*rajat*, *jod or jad oja*, *tho or do*) crop is mostly grown under rain-fed conditions or where irrigation facilities are available. A crop sown very early will make growth before severe winter sets in to an extent that it may be damaged by snow and frost. If sown very late, the tender seedlings would find it hard to make a rapid growth during the spring time. Therefore an appropriate timing is very essential. Wheat is used for bread or chapatis, Lentils (*kerche*) is sown in October and cut in May and June. Coriander seed (*bugo*) is usually sown from the end of September to the middle of November and ripens on low lands in April and on high lands in May. It is sold as a spice. Field peas (*nyar*) are sown in November and harvested in May and June. Other smaller crops are *phapur*, *bathoo*, *cheema*, *kodu*.

Kharif crops are sown from March to middle of July, and reaped from July to September. Crops sown usually in this season are lesser millets, potato and maize. *Chiche*, *chit*, *chay*, *chid*, are sown in July and cut in September and are used usually like rice. Maize (*chhalia*) is grown, as a rule, on *bakhal* land only. *Dankhar* has two varieties, red and white, but the product of both is white. Its leaves, when young, are used as vegetable. The grain is used for bread. Its species *taka* is also used in a similar way. There are several other local crops like, *kodro*, *olgo bras*, *dao* and *chabrya*. *Botang*, a pulse is used as pulse, but sometimes baked into *chapaties*. Potatoes are also cultivated in some areas.

The kitchen vegetables known some hundred years ago were only peas, beans, greens, turnips and some potatoes. But now cabbage, turnips, peas, beans, pumpkin, tomatoes, chillies, radishes, greens, and garlic are some of them which are produced. Tobacco and other drug yielding plants are also being gradually cultivated.

An agricultural exhibition is now organised by district authorities at Kalpa every year for the benefit of the cultivators.

HORTICULTURE

Some forty years ago there was no systematic fruit cultivation, yet a good amount of fruits was available such as apple, apricot, peach, chestnut, and walnut. This shows that the land is endowed by nature with excellent climate and soil for the production of a variety of fruits. With scanty rainfall, this area is most suitable for the growth of dry and temperate fruits. No systematic horticultural work was done in the past, but there were plenty of apples, grapes, and dry fruits. Now a chain of fruit research stations, demonstration orchards, and nurseries have been established to develop horticulture in this region. This development came about only after the formation of Himachal Pradesh. The annual yields of apples, grapes, chestnuts and walnuts has increased considerably. Apples, have recorded about sixteen times increase, whereas all others have shown a hundred per cent increase.

Production of temperate vegetable seeds, such as cabbage, carrot, turnip, sugarbeet etc. have been taken up beyond Tapri. Prior to partition these seeds were produced at Quetta. Now a Research Station in collaboration with the Indian Council of Agricultural Research has been established at Kalpa and a sub-station at Ribba. Its aim is to conduct research on different aspects of seed crops and to encourage and guide the cultivators.

Forests

The total population of this district is wholly dependent on the forests for their day to day needs of timber and firewood.

Livestock depends for fodder and grazing on this forest. For the state, forests were a good source of revenue and even now contribute substantially to the revenues of Himachal Pradesh. Besides producing rains and sustaining fauna, the forests produce a variety of medicinal plants which are a valuable source of income to the people. A sizeable part of the wooded area is covered by *chilgoza* trees producing these edible nuts which fetch a handsome price and form a substantial source of income to the population. These forests were leased out to the British government from 1864 by the state of Bashahr. But now they are managed and developed by the Forest Department of Himachal Pradesh.

A Forest Research Division was created in 1961-62 to conduct research on various forestry problems. Besides grafting of *chilgoza* on *chil*, a *chilgoza* nursery was raised in 1964-65. Several medicinal herbs such as *kark*, *ban*, *kakri*, *kulzira*, *belladona*, *dhoop*, *chireta cirtemisia*, *ephedara* etc. are found in these forests. Due to the rising demand and the high prices offered in the market for the drugs, trial cultivation of a few species was carried out, but was not continued and the active principal contents of those herbs were not fully evaluated. It appears that the cultivation possibilities of drugs are very fair.

For the proper development of agriculture, horticulture, and forestry, the state government has taken several positive steps in this direction. Seeds and chemical fertilizers are made available through several depots on subsidized terms. Chemical fertilizers are not very popular here and hence sludge or human excreta, dung and droppings of domestic animals and in some cases, leaf mould too are used as manure. In the upper regions where there is scarcity of leaf and animal droppings, the people use their night soil as manure. Each family has a latrine in which the night soil is kept for a few months and is then carried to the fields before ploughings. Besides this an age-old practice of mixed cropping, particularly in the *kharif* season still goes on. This method of mixed cropping ensures preservation of fertility of the soil.

Government is providing all possible assistance to farmers for increasing the agricultural produce. The field staff of the Agriculture Department have brought modern techniques of agriculture to the people. Advice along with agricultural implements, fertilizers, insecticides are provided by the state. Financial assistance like loans are also available. *Taccavi* loans are given for several agricultural purposes. The rate of interest varies from four to five per cent in respect of different kinds of loans and recovery is effected in easy instalments, in one to fifteen years. Several other loans are also available for victims of fire and flood, for horticulture and garden colonies, for housing to low income groups, and village housing projects.

Several garden colonies are to be established with the help of technical guidance of the Horticulture Department. Two such colonies have come up at Nachar and Spilo. Village leaders training camps have been organized to acquaint the cultivat ors with the scientific and economic methods of cultivation. Insecticides are supplied to cultivators either free or on fifty per cent cost. Several research centres at Kalpa, Ribba, Giabong, Sharbo, Kilba, and Kamru are functioning for dry fruits, grapes, and *zira* (cummin).

HANDICRAFTS, COTTAGE INDUSTRY AND NATURAL RESOURCES

Kinnaur had been absolutely secluded from the rest of the country, hence people had to depend on their own economy in the past. Every village was self sufficient and people were provided with their necessities among themselves by co-operation. The dry and cold climate forced them to take to a subsidiary occupation of weaving woollen cloth for home consumption and sale, in case of surplus. Shoe (*pono*) making, carpentry, blacksmithy, basketry, oil extraction, distilling, bee keeping were in practice, for the use of every householder engaged in agriculture, trade or keeping livestock. People's devotion to deities inspired them to make metal sculptures, artistic metalwares, musical instruments, fresco painting, and wood-carving in the local temples. Embroidery and artificial flowers were also made by the inhabitants for the purpose of decoration. Earlier travellers of the nineteenth and early twentieth century had made several references to these crafts.

Woollen cloth was the first requirement to cover the body in the cold climate. Livestock, particularly sheep keeping has been the most common thing, which provided wool for the home-spun woollen cloth. The joint family system had helped in the division of labour in the family such as rearing sheep, looking after domestic affairs and women to the spinning work. After obtaining wool from sheep, it is washed and dried. Combing and carding is done by means of wooden combs at home by the ladies. Men and women both spin it by means of a *takli* or spindle. Thus woollen yarn is obtained. This yarn is handed over to an artisan, who spreads it on the ground, passing it round pegs fixed at certain distances, equal to the length of the article to be manufactured. Weaving is done generally by the Harijans, and they are found in every village. Weaving operations are done in winter only as weavers are engaged in agricultural work during the other seasons. Only in exceptional cases such as to execute special orders to cater to the needs of marriages, the weaving may be taken up in other seasons also. Towards the end of the nineteenth century, one Mr. Schreve, a Moravian Missionary had introduced a handloom from Europe and had taught people to make blankets. The age-old method of weaving is said to be uneconomic and products are of low quality resulting in small income. There were 408 cottage industry establishments in 1961. The main articles manufactured are *pattus* (Shawls), *dohris* (woollen garment like sari), *chhanli* (blankets), *pattis* (tweeds), and mufflers. These articles were primarily meant for domestic consumption or made against order for which the material was supplied by the customers and the craftsmen were paid labour charges only. Surplus articles were locally bartered. Only neglected pieces were displayed at *Lavi* fair at Rampur. Previously the products were exported to Tibet. Now they have ready market at Simla and in the plains. Due to the development of transport, goods from Kinnaur are now readily available at Simla.

Pono, the original crude footwear was made by Chamangs for the use of the local people. Raw materials like wool and leather were locally available. The lower portion of the shoe was of leather and the upper was of wool. The stitching was done by leather thread. They were either plain or embroidered.

Embroidered shoes were known as *tapru-se-balzam-pono*. They are still made but now, the shoes from plains are in great demand. Now Chamangs have started manufacturing leather shoes.

Wooden parts of the agricultural implements are made by carpenters. The ores community is engaged in wood work. They construct houses and temples alike. A carpenter is needed in every village. He manufactures wooden bowls and dishes for domestic use. There are several craftsmen at Kalpa, Thangi, Rarang, Sapni, Batseri, Shaung, Bari and Bhabha who make wooden items of utensils like *zom*, *zwach*, *gam* and *chaoring*. Vessels of juniper wood made at Sangnam and Ropa were exported to Ladakh in earlier times. Now they are available at Simla government emporium as show pieces.

Blacksmiths are the members of Domang community. They make all kinds of agricultural implements and musical instruments for local temple. The chief articles prepared by them are *bham* (a large kettle-drum), *nagari-cho* (pair of kettle drums), ronsheeng (*hornpipe*) and *dumko* (a vessel used for distillation of local wine). Generally the workshop is located on the ground floor, but in the case of single storey houses, it is in the front of the house, in the open space covered by walls on three sides. Skin bellows are fixed to the hearth. The whole process of melting, purifying and welding is done to produce a finished product, an agricultural implement or an utensil or a hornpipe.

Basket-weaving is done by Chamangs at places where raw materials like *rajial*, *tako kas'yang* and *teg-shing* are available. *koting* or *kilta* (conical basket), *chhatoch* (basket with handle) and *changor* (basket) are generally made, usually in winter months. *Koting* or *kilta* type basket is in great demand for carrying loads on the back. Other baskets are used for household needs. In the plains, bamboo or *arhur* sticks are used. The process of manufacturing is also similar.

Oil extraction is done in almost every household. It is extracted from the kernel of *chul*, *reg* and *ker* by the traditional method. Kernels of apricot, peaches, and walnuts are taken out and dried in the sun for two to three days. These dried kernels then are put into a *kaniry* (stone mortar) and pounded repeatedly with the pestle to a thick paste. This paste is boiled on high heat in a cauldron, and later small cakes are made out of the heated paste. Now each cake is placed on the edge of the mortar and pressed hard so that the oil starts oozing out of the cake and running into the mortar. This is kept in metal or earthern pots locally known as *gagri and pathu.* In the plains the oil seed is pressed in a *kohlu* driven by an ox or man, and oil comes out of it. Like in Kinnaur the oil-cakes or (*khali*) are used as cattle food. Now oil mills driven by power have come up. One such mill is located at Kalpa.

In Kinnaur liquor popularly known as *ghanti* is prepared for household use by almost evrey family. Both distillation and fermentation methods are used. These are prepared by all kinds of fruits and grains. *Arak* or *rakh* or *phasur* or *anguri* are prepared by distillation and *ningo*, *chhang* or *shudung* and *boga* by fermentation. Every householder is granted a licence and he is required to pay rupees twenty five per year to the state.

Arak is distilled from *tag* or *nanga jo* (Barley). Barley after putting in sufficient water is heated and then it is processed with *phup*. *Phup* is a special ingredient or fermenting agent brought from places like Chango and Rampur. It is very cheap about thirty or forty tablets are available for a rupee. Barley with *phup* powder mixed is kept in a closed bag for fermentation. After it is fermented distillation is done by heating with water in an airtight vessel and the steam water is collected in a pot. The process of distillation continues so long as the liquor has been distilled to a desired quantity.

Dakhang pashur (*Anguri*) is prepared out of grapes (*dukhang*) by distillation, mostly near about the village Ribba.

Ningo chhang and *boja* are also prepared from Barley. The fermented barley is kept in a *jama* pot with a whole in the bottom. To produce *ningo* a little water about two bottles is poured into the *jama*. This vessel is kept for twenty five days or so. After this period the lower hole is opened and water percolates into a pot. When the above quantity of water is out, the hole is closed and the best quality called *ningo* is ready. In the same *jama* when *ningo* is taken out some water about 25mm above the grain level is kept and after about twelve hours this water is taken out from the hole in the same fashion as *ningo*. Now this water is known as *chhang*. The remaining grain is given to cattle for eating.

Boja is also prepared from barley. Fermented grains are poured into a copper pot and rubbed with the hands. It is rubbed so hard that the mixture of grains and water is turned into a thin paste. This paste is filtered by means of a sieve. The liquid thus obtained is served to the people as *boja*.

Home made liquor is stored in the house for daily use or for festivals and ceremonies. Some times, it was sold for rupees three a bottle.

Bee keeping is said to be very primitive. The hive or *yang* or *mokhar* consists of a box or a stone slab with a small hole. This hole opens into the wall of the house. The hive is set at a sunny point. The outer side of the box or the hollow in the wall serves as an exit and entrance and can be opened at will without disturbing the bees. This hive is opened only twice a year, once in March and again in September, when it is supposed that all the *wass* (honey) is extracted. Bees are smoked out through the entrance hole with the help of smoking wood or rag or a dung whichever is available. After extraction the honey is purified by filtering. Bee hives yield wax also which is used. This honey and wax are then used. Sometimes the products are sold also to supplement the family income.

Artificial Chambaka flowers are made by almost every lady in Kinnaur. Thin paper like material available in Sirmur, Mandi, and Bilaspur districts are used for this purpose. This material is given the shape of semi-circle. A few pieces are tied together with a straw so as to make a bunch of Chambaka. These are used on caps by almost everyone in the region.

Metalware and ornaments are made by Domangs, who are scattered all over the district. There is at least one in every village engaged in this part-time occupation. They do solid and hollow casting, prepared engraved metalware, ornaments and traditional musical instruments on order. These crafts have been in vogue since time immemorial. It is said that these crafts have been imported from Tibet from Tibetan mastercraftsmen. These same artisans when required to do so produce gold ornaments like *murki* (earring), *balu* (nose ring), *chander sen har* (necklace) and *long* nose ornaments) on order.

The artistic taste of the people can be judged from the many local temples which contain wall paintings. In Lippa, Rispa, and Nesang here are a few painters well known for their paintings on fresh plaster, cloth, and paper in the Tibetan style. Paper is made of *ziko* bark. The design used by painters are unequally local and religious.

Natural resources were available in plenty to the people of Kinnaur. This is evident from the following passage :

> There are several iron mines in Bussahir which are worked by the inhabitants of the country, who appear in good circumstances. Some of the galleries extend half a mile horizontally into the side of the mountain, but they are not above three or four feet wide. They use no perpendicular shifts, so that the miners are obliged to work by torch-light.
>
> Nawur may be considered the principal mining state and there the iron sells for 12 seers per rupee. Bussahir government levies a duty of 3/4th of an anna upon a load (as much as can be carried by a man), which is from 40 to 50 seers, or nearly one Cwt. The place of rendezvous for the sale or barter of the iron is Sisco, a town in the Protected Sikh plains about 34 miles from Subathoo, and there the mountaineer either receives cash or produce of the plains, such as goor, grain, tobacco, clothes, sugar etc. etc. for his iron.
>
> None of the precious metals are found in Kunawur or any part of Bussahir. If a vein of metal was discovered the constitution of the mountains, composed of greiss, granite and other most hard rocks, and the impracticable nature of the country, would entirely preclude the possibility of working it to any advantage. Copper, which always contains more or less gold, is found in one part of Kunawur. The mine was worked for a short time by a miner from Cooldo, who abandoned it as soon as the ore on the surface was exhausted.
>
> A gold mine was discovered a few years ago in the Tartar dominions adjoining Bussahir, but for what reason it was immediately shut up by orders from the 'Mahacheen' is not distinctly known.[6]

James Baillie Fraser has also reported about the existence of copper and iron. Gerard observed that gold dust was sometimes found in small quantities, at a sandy place below Morang in Sutlej, which is commonly called *Zung-Tee*. The first word means gold, and the latter water. This occurrence, however, seems too poor for commercial exploitation. Similarly, copper mining in Ropa valley is also reported by Gerard.[7]

In Kalpa tahsil near Purbani iron pyrites are reported. It seems these were exploited by the people in earlier times. Near 'Chugaon in large pegmative vein, nearly fifty metres thick, silver is reported to occur and the evidences of old working have been recorded'. Evidence of mica is also available. 'States are quarried at Bhabha in Nachar tahsil, at Kalpa, Khangi, Rogi, and Yuwarangi in Kalpa tahsil; at Barua, Chasu Kilba and Rokti in Sangla tahsil and at Rarang in Puh tehsil.[8]

There are possibilities that the minerals like copper, iron pyrities, gold, silver, mica, china-clay, gypsum, lime stones and dolomites can be discovered and the resources thus available be used commercially.

Himachal Pradesh after the establishment of the district has come out in a big way to develop and provide training facilities to the different cottage industries of the area. The possibilities of the hydro-electric power are being investigated and a hydro-electric power station at Nogli is established. Under the Mehbar micro hydel scheme, two generators of 100 Kw each were also commissioned. Several villages have already been electrified.

Many production-cum-training centres are established for various crafts. Blanket and tweed production centre at Nachar was established in 1961. This centre produces blankets, tweeds, serge, *pattis* and *pattus*. A centre at Spiloo was also established in 1961 for the production of *Gudma*, *Namda* and *Chugdon*. Wood working centre at Dune was opened in 1962 for the manufacture of furniture for use in various offices in the district. A carpet and drugget production centre at Sangla was started in 1961. It manufactures carpets, druggets and *dohris* for which there is a large local demand. The leather centre at Kilba was converted into a production centre due to lack of trainees and is now manufacturing shoes, chappals and attache cases. These products are sold locally. Government Craft Institute Kalpa was established in 1962. It imparts training in crafts like tailoring, cutting, knitting, and embroidery. In 1968 this unit was merged in Cluster Centre, Duni now known as Rural Industrial Training Institute. After merging several small units a government Industrial Extension Centre (Textile) has come up at Nachar. The Hosiery Centre, Morang imparts training in making sweaters, socks and jerseys, but it has been closed for want of raw material. An oil pressing centre was established at Nachar in 1964. The articles produced in the above centres are mostly consumed by the local population and sometimes placed in the emporia at Kalpa, Simla and various other places in the state for sale.

TRADE AND COMMERCE

In Kinnaur trade had a great role to play. It was most essential, to some extent even compulsory. Kennedy in 1824 has remarked 'In Kunawar the people chiefly subsist by trade . . . . The prosperity of Bussahir mainly depends upon this trade'.[9] In the past Kinnaura traders used to trade not only with near-by Tibet and Ladakh, but they used to visit far off places like Central Tibet, Yarkand, Kashmir, Lahore, Delhi, Calcutta, Bhutan, Sikkim and Nepal. James Baillie Fraser remarked about them in 1814 that, 'Indeed they are almost exclusively the commercial couriers between Hindustan and Tartary, as also between Tartary and Kashmir, frequenting the routes from Leh in Ladakh, to Lhasa and Degrucha and Nepal on trading speculations.[10]

Soon after the Gurkha War in 1815, the British East India Company made enquiries about the trade potential of the hill regions, particularly of the Bashahr state. Captain R. Roso Assistant Agent to Governor General deputed one Munshi Karimuddin to Garo and Ladakh for the purpose. He submitted a remarkable report to the government detailing routes, distances, commodities of trade with rates of selling at Garo in Tibet and Lavi fair in Rampur (See Appendix IX). Traders from Kinnaur used to visit Tibet in the month of May and June and return in the months of June and July, whereas traders from Tibet came in the months of October and

November and returned in the months of November and December. They travelled in batches of twenty to twenty-five carrying provisions with them as the Garo in Tibet had no permanent habitation, and nothing was produced there. Tibetan traders used to carry a red blanket which was their dress at Garo. Trade items were salt, *sohaga*, goats' wool, *nerbussee* (a drug), shawl wool producing goats, and tea.

Salt was bartered for *drug* (a sort of raisin) and wheat at equal quantities and was sold at Rampur one seer for 4 seers of rice or 8 seers of *mundooa attah* or 16 to 20 seers for a rupee. Sohaga was bartered at the rate of 5 seers for 4 seers of *drug* and was sold at Rampur 7 to 8 seers per rupee. *Nerbussee* or (*nurbees*) (a drug) was brought from Nepal and was sold to Kinnauras at 3 pieces for a rupee. *Goats* producing shawl wool were purchased at the rate of two rupees each and sold for 3 rupees in Kinnaur. A box of tea of 10 seers cost 10 rupees at Garo.

Trade with Ladakh was also much in progress. Silver was purchased at Ladakh and was sold in Bashahr for the Moohamud Shahee rupee. The comparative value of these two was 4 to 4, a fifth was thus added to its price. *Namda* with border of the size 5 *hauth* × 3 *hauth* was purchased for 4 to 5 rupees and was sold at Rampur for 7 or 8 rupees. *Saffron* was purchased at rupees ten for a seer and was sold at two rupees per tola. The chief articles of export were *heeng* (*asofoetida*) *zeera* (*cummin*), *bhojeputtra*, opium, charas, *mohru* (a poisonous drug), iron, white puckes (woollen stuff), *sucklad* (woollen stuff), mandi rock salt etc.

There used to be three fairs at Rampur in the month of December, October, and June every year. At these fairs everything was bought and sold. From tobacco to swords, everything from the plains of India was available.[11]

At Lavi fair, 'Tea from China and bars of silver, stamped by authority, from Yarkand, toys from Russia, may occasionally be had at the Rampur Fair. Mandarin Chosticks, cups and saucers are seen in Kunawar. Dresses of men of rank, some of them very costly, silks, satins of very beautiful patterns and te ture come also to the same fair'.[12]

All the travellers in the course of nineteenth and early twentieth centuries like, Fraser, Gerard and Rahul Sankrityayan spoke highly of this Kinnaura trade. Prosperity of the people was mainly due to trade. Several items of trade were procured from the plains and were exported to Tibet and Ladakh. Among the local products the chief items were, timber, *neoza, kadu, patish, zira, dhoop, gudmas* (blankets), *pattis*, yak-tails etc.

MONEY AND BANKING

Money was not much in circulation in this region In the course of his travels in the early nineteenth century Gerard had noted that, 'The coins are rupees, both the common one of Hindustan, the Moohamud Shahee, five or six per cent better; and one current on the hills, only half the value of the former. The *trimasha* or *paolee*, a silver piece of four pence, and a copper coin, called *dubwa* in value from half penny to a penny according to the size . . . . Commerce is chiefly carried on barter.[13] In course of time the British system of coinage was introduced, and now decimal currency is in use. In early times the method of exchange popular and handy was the barter system. In course of time the Zamindar *sahukar* or money lender appeared. Earlier the priest of the local deity who gave money from the temple was the only source for getting a loan. In this way the temple under the priest served as a bank. These zamindars and *sahukars* started lending on the security of land or jewellery. Side by side the local deity also functioned as bank advancing loan of money and grain on stipulated rate. Here the guarantee was only a solemn oath before the deity. In practice nobody had ever cheated the deity. In course of time the demand increased for various reasons, such as construction of houses, for marriages, for investment, in trade and commerce etc. This demand for money coupled with the increase of population went up several times than it used to be. Thus demand and necessity had increased the rural indebtedness of the people who often carried this burden for several generations. The rate of interest had been low as compared to other rural regions of India. Perhaps it has a direct connection with the capacity to repay. The population here by and large was quite poor.

The indigenous economic pattern continued till very late. Himachal Pradesh was formed and the new border district of Kinnaur came into existence in 1960. The credit facilities came into existence in the form of Co-operative Credit Society, Agricultural Multipurpose Society, Service Cooperative, Multipurpose Cooperative Society, Cooperative Bank, Land Mortgage Bank etc. In the meantime the India-China Trade Agreement of 1954 expired on 29 April 1962. This led to the closure of trade with Tibet. Not only did traders become idle but livestock was also denied grazing facilities in Tibet. This created a problem for both men and animals. The fodder problem was solved by sharing the existing the grazing land in Himachal Pradesh and the human problem was tackled by the state by starting a cooperative marketing and supply federation through which traders obtained loans for their business etc. The two societies have in substantial measure rehabilitated most of the traditional traders. The Rang Foch Simant (muleteers) Cooperative Society which with headquarters at Puh has been transporting supplies for the local people as well as the army, has earned a substantial income which has been fairly distributed among all the members who were former traders and owned a number of pack animals. Similarly, the Kinnaur District Cooperative Transport Society has been transporting goods to the remote areas of Kinnaur in jeeps, and with the widening of the roads has recently started operating a truck to enable larger stock to be transported. The petty traders, who either do not own animals, or do not have enough land or other means of livelihood to make up for the lost trade with Tibet are taking up alternative business, such as opening of shops, owing of improved breeds of sheep etc. Grant of interest-free loans is greatly helping them in solving their resettlement problems. In the wake of the Chinese aggression in 1962 the construction of roads on a large scale has provided employment to several people. Construction of several district office buildings and furnishing have also provided jobs to the people. All other state agencies in several spheres are functioning now.

New trade centres are developing at Kalpa, Tapri, and Karchham. Several retail centres are also coming up at Kalpa, Tapri, Pawari, Shontong, Puh, Aran Jhula, Chauhra, Wangtu, Nachar, Sangla and Karchham. Besides these, several small shops and hutments all along the National Highway and other feeder roads have been established. Several sudden developments have not only improved the economy of the region but have influenced the confidence of the hardy people of Kinnaur.

> For the population of Kinnaur from very early times, money meant ornaments and hoarded spice.
>
> Ornaments were a form of fixed deposit. The deposit was mostly in silver and gold was but a small fraction. Exchanges were mostly effected by barter and wages paid in kind. It was a dependably self-sufficient economy. The cereals, except part of wheat and all the rice (save for the little rice produced at Kilba village) came from the domestic or local land; milk, ghee, butter, curd, butter-milk and meat from the domestic or local livestock, honey from the domestic or local beehives, domesticated or wild; and clothes and shoes from the wool, the goat hair and the hide of one's own livestock or the livestock of the locality or the neighbourhood. And for even tobacco and drinks one did not have to be extraneously dependent. It was a home spun economy, neither luxurious nor extravagant, but steady and secure, even if not always abundant.
>
> The conversion of the land revenue from kind and service to cash payments started the thin end of the wedge of the cash nexus. The currency of coins never looked back. In course of time paper money took over here as elsewhere. Imperceptibly to begin with, but apparently later, the expansion of the cash nexus went or changing the basic quality of the economy. With the sudden outburst of developmental activities following the Five Year Plans, especially after the creation of the border district, money has been pouring in as never before, and, in the wake of the rapidly expended means of communication, market for cash crops, . . . undreamt of before, have come within reach. The barter economy is dead and cremated and with it the home spun and steady quality thereof. The economy is still in a flux, a melting pot and a confusion, bordering at times on chaos.[14]

REFERENCES

1. From Captain C.P. Kennedy, Assistant Deputy Superintendent, Sikh and Hill States, to Lieutenant W. Murray, Deputy Superintendent, Sikh and Hill States dated Sabathu, 6 July 1824, *Punjab Government Records: Delhi Residency and Agency, 1807-1858* (Lahore, 1911), Vol. I, pp. 255-99.

2. Directorate of Economics and Statistics, Government of Himachal Pradesh, *Statistical Abstract, District Kinnaur, 1972-73*
3. M.D. Mamgain, ed., *Himachal Pradesh State Gazetteers: Kinnaur* (Ambala, 1971), p. 146.
4. Captain Alexander Gerard, *An Account of Koonawar in the Himalayas* (London, 1841), pp. 63-66.
5. Mamgain, ed., n. 3, p. 129.
6. *Kennedy Report*, 1824.
7. Gerard, n. 3, pp. 27, 273.
8. Mamgain. ed., n. 3, p. 169.
9. *Kennedy Report*, 1824.
10. James Baillie Fraser, *Journal of a Tour Through Parts of Snowy Ranges of the Himalayan Mountain and to the Sources of the Rivers of Jamuna and Ganges* (London, 1820), pp. 274-5.
11. N.A.I. Foreign Department, *Secret*, 27 October 1815, Nos. 39-42. *Translation* of a paper respecting the Trade between *Bussahir and China* by Moonshi Currumooddeen.
12. *Kennedy Report*, 1824, See also paras 68 to 70 for trade.
13. Gerard, n. 3, p. 86.
14. T.S. Negi, *Scheduled Tribes of Himachal Pradesh, A Profile* (Simla, 1976) pp. 24-25. Material in this chapter as elsewhere in this work was collected after extensive tour of the region and after prolonged discussion with the people and Government officials.

5

Political Organization and Administration

EXECUTIVE MACHINERY

It is not possible to form any idea of the administration of ancient Kinner Desh. The nomadic pastoral people of the region must have gradually settled in the small villages with their local deities. Amongst them the physically most powerful, must have grown to the position of Thakker or Thakur. Similarly, in course of time the most powerful amongst these Thakurs must have emerged as all powerful kings or Rajas. Kamru Thakur after subjugating several thakurs of the Kinnaur region emerged as a Raja of Bashahr. Thus an autocratic polity came into being. It appears that there has always been a certain amount of independence enjoyed by the people who used to organize themselves without much interference from the Raja. Every village had a body which included *Mukhiya*, a *Char*, a *Halmandi* and a *Toknya*. Halmandi and Toknya belonged to the Scheduled Caste community. Besides these there used to be a kind of Panchayat consisting of two or three noblemen. The main function of the Panchayat was to collect land revenue and settle disputes. A state official (daroga) would visit every year or two to settle major disputes, if any. This system was functioning under a Sarhaddi Wazir, who was in charge of the frontier and was invested with considerable independent authority. His office was not hereditary. The ablest man available in the area was appointed as a Sarhaddi Wazir. He was practically given a free hand as long as he maintained order.

In order to maintain a direct link with the people, the Raja in ancient times used to send out his men whose main function was to report about the criminals and crimes. They also reported about the taxes levied upon the people, i.e. to say who was taxed beyond his means and who had not been taxed properly. These men were called *Churiahs* and *Hazrees*. These were secret service people equivalent to modern intelligence and revenue intelligence departments. The payment system of these men was the same as was prevalent elsewhere in the country. They used to get rent free lands and no cash payment was made to them. This system must have been prevalent before the Gurkha War. The Raja must have been in full control of the state authority and the Wazirs kept at their legitimate positions as counsellors. But during the minority of Raja Mahendra Singh and after that the three hereditary Wazirs became very powerful. Kennedy in 1824 observed that :

> The Government of Bussahir consists of the Raja and three hereditary Wazirs. I beg leave to notice the evil that arises to these districts where these Wazirs are hereditary.
>
> As soon as they obtain an influence over the Chief by indulging in his propensities, intoxicating his senses with drugs inducing a degree of mental imbecility, then commences the abuse of men dressed in a little brief authority; money is their only god and they do not hesitate to give ample scope to their rapacious and tyrannical propensities.
>
> The present Raja is 14 years of age. He was placed upon the Musnad in 1821. Only a few weeks after his inauguration his mother, who had the guardianship of her son in his minority, was burnt to death, by accident or design is not quite ascertained, but it is generally supposed the latter, and at the instigation of the Wazeers, in order that they might be better able to continue the system they had pursued of keeping the reins of Government in their own hands.
>
> The education of the Raja has been grossly neglected and already he has the appearance of being an opium-eater. The Wazeers allow but a very trifling sum of revenue which they have charge of to be appropriated to the expense of the Raja's state or comfort.[1]

Thus the authority of the Wazirs was complete till the death of Raja Mahendra Singh. After his death his eleven year old son succeeded to the Gaddi. The old game of hereditary Wazirs continued. So much was the powers of the Wazirs that 'the relations with the Raja and Wazirs, which takes large share in the management of the state, remains somewhat strained'[2] The situation deteriorated gradually. His sister 'Deri Sahaba' had also a considerable influence on him. The vice of drunkenness which he had long been addicted to begin to grow steadily. So much so that it was remarked that 'the Raja is given to seasons of intemperance, and is often for a week or ten days hopelessly intoxicated. In such conditions Sanads are issued which he repudiates when in senses'.[3] This state of affairs became intolerable very soon. At the same time the oppression of the Wazirs also grew steadily. Ultimately the Raja was prevailed upon to hand over the administration to his only son Tikka Raghunath Singh in 1887. Raghunath Singh with the Wazirs as counsellors, not as dictators, began to manage the state administration for the welfare of the subjects. Many of the subjects under the old regime had been so oppressed as to forsake their homes and lands and had emigrated; they came back to their homes in the new regime.

Tikka Raghunath Singh took serious pains and organized the revenue and civil administration on somewhat progressive lines. Under the influence of corrupt Wazirs the old Panchayat were also corrupted and they were abolished in 1895. Bashahr state was divided into three tahsils for the purpose of administration. The *tehsils* were subdivided into *parganas*; parganas into *ghoris* and *ghoris* into villages. Chini tahsil, i.e. Kinnaur was divided into five parganas which stood further subdivided into fourteen *ghoris*, and these into sixty-three villages. Each pargana had attached to it a fort located at a central place with an establishment according to its needs. Every pargana was under a headman called *dashongi*. This official was paid from the revenues collected under him. He used to get one per cent of the revenue. Each *ghori* was under the charge of a *lambardar* who was getting three per cent of the revenue of his charge. Each village was under the charge of a *char* (watchman). His percentage in the revenue was only one per cent.

Tikka Raghunath Singh died on 2 February 1898. Till his death he was head of the administration in matters of Civil and Revenue. After his death Rai Saheb Mangat Rai was appointed as manager in 1898 with same powers. He was later succeeded by W.H. Emerson, I.C.S. and continued till Raja Padam Singh assumed powers on 30 November 1917.[4]

During the period of the managership, the manager was at the apex of the state administration and he exercised full powers as the Raja used to in the matters of civil, revenue and criminal cases; and only sentences of death required the confirmation of the Superintendent, Simla Hill States, Chini Tahsil, modern Kinnaur was under a Tahsildar. He had III class Civil and Criminal powers. At tahsil level there was a *kanungo* with six *patvaris*. There used to be a small sub-treasury also.

JUDICIARY

CRIME AND PUNISHMENT

Regarding crime Kennedy observed in 1824 that, 'In no part of the protected dominions, and I may give a wider scope and say the world, is there less crime known'. Disputes if any were settled by arbitration. The crime situation was not different in 1841 when Gerard visited the area. Even in the second half of nineteenth century one comes across observation like 'Fortunately the country wants very little governing'.[5] 'The people of Kanawar and Bushahr seem to be a simple law abiding lot and as long as they can obtain some sort of settlement of their disputes by Wazirs they are content'.[6] In the first half of the present century also the crime situation was not very much different. The honesty of the Kinnauries is proverbial even now. Dr. Van Dar Sleen who visited Kinnaur in 1929 observed about the crime and character of the people thus :

> There are no cases of theft except the theft of women, and, as everyone knows, all is fair in love and war. The penalty for such misdemeanours is always decided by the votes of the neighbours, and is as a rule paid promptly. Neither are there any murders, because if you happen to kill anybody in a fit of temper or drunkenness, that is a misfortune, not a crime. One source of frequent squabbles is the right to land and crops. Water rights too are a special bone of

> contention, such as who may or may not tap the supplies of certain springs. Lucky mortals who are troubled neither with the existence of a legal code nor with lawyers. Here law is simply law, and swift at that. Whether it be always justice I have my doubts.[7]

Kinnaur is no more an isolated area to the extent it was earlier. With the advance of modern civilization the crime situation has also changed slightly. Despite changes heinous crimes do not exist and the nature of the offences committed include theft, house breaking, offences under Excise Law and the Arms Act. Even in these crimes the incidence is negligible. The slight increase in theft and house breaking cases can be attributed to the general poverty of the people.

Since the incidence of crime was negligible there was no need of police force, and no police station was established till very recently. In 1910 there were twelve constables and a sergeant of the Punjab Police stationed at Rampur probably for the security of the English I.C.S. manager. Serious crimes were reported by the tehsildar and were investigated by *daffadars*. Just before independence a police station was established at Rampur under one sub-inspector with two head constables and eight constables. After the formation of Himachal Pradesh one police station at Chini was opened under a sub-inspector. The total force including the treasury guards was four head constables and eighteen constables. After the creation of the district in 1960 the police force was properly organized and now the head of the police organization is a superintendent of police as elsewhere in the districts; now there are several police stations at least one in each tehsil headquarters. Besides police stations there are eleven check-posts for checking passports and inner line permits at several places. There are also eighteen check-posts of bridge guards to protect the bridges and control the traffic movements on the National Highway No. 22 the Hindustan-Tibet Road and other important road links. Anti-corruption Unit and Home Guards are also established now. Besides these there are several Intelligence Bureau men and Indo-Tibetan Border security men.

Jails and courts are the essential parts of an area where the incidence of crime is more, but for an area where there were no crimes, there was no utility of these departments of the civilized world. Still the system prevalent was that :

> The punishments in use are for murder execution and for all other offences, corporal punishments and transmission or banishment from the country. . . . There are no prisons, and the Chieftains are often at loss in what manner to punish where the crime does not come up to murder.
> The Chieftains appear to administer criminal justice with peculiar leniency. Fines are common for misdemeanours and not unfrequently levied without much regard to justice. However the custom of the ages is in favour of its toleration, and complaints but seldom made of its infliction.[8]

In the state of Bashahr there were two prisons, one outside the Kinnaur region at Rampur, and the other was located at Kamru, during the time Rahul Sankrityayan went in 1948. The prison at Kamru was a well-like jail in which prisoners were made to descend and dwell. Bread and water was passed on to them through a rope.[9] In subsequent years prisoners were sent to Solan or Nahan jails. A sub-jail was opened at Kalpa on 1 May 1963, but convicts are generally sent to Solan or Nahan jails even now.

The court of the Raja was the only court of appeal. Till 1887 hereditary wazirs were supreme, but after the assumption of power by Tikka Raghunath Singh, he took over all powers of justice. After his death Rai Saheb Mangat Ram took over as a manager and all powers were passed on to him, and later on to an English manager till the succession of Raja Padam Singh to the gaddi in 1914. But the total power went to him in 1917 and it continued to vest in him till his death. The laws of British India were not applicable in Bashahr state. Only those laws which were suitable to the region were introduced by an executive order from time to time.

After the formation of district in 1960 the usual civil and criminal courts were opened. Sessions Court is located at Simla. All the divisions and sub-divisions have sub-judge courts now. In the matter of criminal courts, besides the District Magistrate at Kalpa there are several courts in every tahsil and sub-tehsils under tehsildars and naib-tehsildars all over the district. Appellate courts besides the District Magistrate, Kalpa are

Sessions Court, Simla and Himachal Pradesh High Court, Simla.

Though ancient Panchayats were abolished by Tikka Raghunath Singh in the last decade of the last century, they have now been revived under the Himachal Pradesh Panchayat Raj Act of 1968. Under this act several *Nyaya* Panchayats are established. They derive their powers from the said Act, suitable to the special local conditions. The term of each *Nyaya* Panchayat is of three years. These *Nyaya* Panchayats are empowered to impose a fine not to exceed by rupees one hundred. They cannot sentence anyone for imprisonment in any form. In Civil cases, except with the agreement of the parties, *Nyaya* Panchayats cannot impose a fine more than one hundred rupees. The *gram sabha* is the limit of the jurisdiction. The appeals against the *Nyaya* Panchayats can be made to sub-divisional judge or the sub-divisional officer, in criminal and civil suits. These *Nyaya* Panchayats are supervised by a well knit system of bureaucracy headed by Director, Deputy and Assistant Director followed by a District Panchayat Officer and an Inspector.[10]

REVENUE ADMINISTRATION

The circulation of coins was very scarce in Kinnaur. Trade and exchange were carried only on the basis of the barter system. The Raja used to share the profits of any kind which his subjects made from their hard labour. Artisans had to give a part of their produce; for instance, bamboo workers supplied baskets; potters gave earthen pots, charcoal burners provided fuel, *kolis* gave ropes or shoes, and so on. This principle was uniformly applied to all products of the country such as animals, minerals or vegetables. The state took a share of the iron smelted in certain parganas and claimed the main portion of the gold dust obtained from the River Sutlej. Imports were levied on every kind of agricultural produce. They had to give wine made from their grapes, grain, honey, or the fruit of wild trees. A quota was taken of the oil processed from the kernels of wild apricots, while the men of certain parganas were held responsible for the collection of honey from the forests. So long as the state was under a strong ruler, the burdens were equitably distributed and the principles underlying their imposition were in harmony with the people's conception of their duties towards the Raja. Kinner region was always favoured by the Bashahr rulers. Therefore, the people were lightly assessed and were happy and contented. During the Gurkha War the Raja and Rani took refuge here and were cordially treated. The hardy Kinnauras were almost the only soldiers who stood with the Raja till the last and by destroying the bridges, and throwing other obstacles in the way of the Gurkha army succeeded in warding off their occupation of this area. This was an added cause for the favoured treatment of the region. But soon after the accession of Raja Mahendra Singh in 1815, the situation began to deteriorate. The power of the hereditary wazirs grew gradually.

During the strong Raja's regime also the value of the articles collected by the officials as their share amounted to twice or even *three times* as much as collected for the state as *karahads* or *habubs*. But the number of *karahads*, or levies in kind, tended continually to increase, during the ascendancy of Wazirs.

> A list prepared by Sham Lal at the middle of last century contained as many as sixty-four items. Some of these were purely local, while the distinction between others appeared to be nominal. The incidence of particular taxes or levies varied between wide limits according to the circumstances of the *Malguzar*. These as collectors of the numerous imports took more for themselves than for the state. They employed a host of unscrupulous subordinates, who robbed the people in all directions. The only remedy the latter had was to organise a periodical *Dum*. This Dum included an effective method of 'peaceful picketing'.[11]

Begar (Forced labour) system was more or less universal. The ruler used to share all the assets of the people. This was applied to such an extent that a man's physical strength was considered to belong to the Raja. It was obligatory on every household to supply an able-bodied man to labour for the state for half the year. Large households and joint families were encouraged by the state. The enlightened British rulers, when they went to the region, debated this issue at great length. But the consensus that emerged was that the system could not be dispensed with outright. In concurring with officials Ochterlony suggested that some rates be fixed, may be four rupees per month only.[12] This, however, remained on paper only and the *Sanad* concluded with Raja Mahendra Singh gave official recognition to the hated system of *Begar* (Forced labour).

Under such a situation the people were expecting some improvement in the administration with the change of the ruler. But it was very sad that at the succession of Raja Shamsher Singh his age was only eleven years and the ruling clique of Wazirs continued. Having no way out of the exploitation of the Wazirs, the cultivators and artisans made a complaint to the Deputy Commissioner. With the consent of the population he issued an order for the introduction of the new settlement of Chini Tehsil of Bashahr State. Under an able officer Sham Lal, the first summary settlement of Chini tehsil, that is sixty three villages of modern Kinnaur district was conducted in 1851 and 1852. The assessment was based on the quantity of land and the seed used in it. After taking into account the above two factors the *khasra* ahd *khewat* were prepared, and the rate of the land revenue and other on the basis of twenty five per cent of the goods produced was fixed. Eighteen *karahads*, i.e. *habubs* (cesses) which then existed, were abolished. The gross revenue of the region was assessed at Rupees nineteen thousand three hundred and nineteen (19,319), Rs. 16,655 land revenue and Rs. 2,664 from cesses.

Sham Lal's settlement could work only for two years. The fact was that there was a shortage of currency and cash assessment with the abolition of the *karahads* with limited opportunities for exploiting the people was against the interest of the wazirs and jagirdars and they got it cancelled. From 1853 to 1928 there have been in all six settlements within a space of seventy-five years. This lack of uniformity of settlements led to the constant struggles between Raja Shamsher Singh and his hereditary Wazirs till the taking over of the state powers by Tikka Raghunath Singh in 1887.

Lord W. Hay's assessment of 1856 was done in a conference held at Simla in which the then Raja Shamsher Singh and representatives of the people were present. The twenty-five per cent assessment of gross produce fixed by Sham Lal was reduced to fifteen per cent. This assessment amounted to

Rs. 10,797 as land revenue and Rs. 1,728 for cesses. The total came to rupees twelve thousand five hundred and twenty five (Rs. 12,525).

Though the settlement by Lord Hay was extremely light, yet it was not in the interest of the Wazirs. The money assessment was not in the interest of the Wazirs. This money settlement had stopped their indirect gains, which used to be enormous and exorbitant. The Raja was definitely in favour of money assessment but he encouraged the Wazirs against it. While the British Government was busy suppressing the Indian Revolt in 1857, the Raja issued orders in favour of the old method of collection of revenue in kind. He justified his action saying that he had done this in order to please the gods, as the state then was under the grip of a Cholera epidemic.

For sometime both systems were in operation resulting in a great confusion in accounts. Sometime later the Raja again reverted to money settlement but as it was not in the interest of Wazirs they began to resist it. Thus there was a major dispute betweeen the Raja and his Wazirs. Ultimately the question was brought to the notice of the Superintendent of Simla Hill State. He on the advice of Government told Raja not to have confrontation with the Wazirs. Thus again the old system of collection of revenue in kind was resorted to. In fact, the chances of the exploitation of the population in the money assessment were less.

Several zamindars of the region submitted a petition to the British Government containing eleven demands in 1858. One of the demand was that the assessment should be reverted to the old system. One Mr. Barnes was deputed to settle the matter. While narrating the history of the settlement he reported that,

> at one time Sham Lal's settlement has been set aside and a partial return made to the old system, with the difference that the tribute to the British Government of Rs. 15,000 a year, which is levied ratably from all the *zamindars*, was increased 50 per cent or to Rs. 22,500 a year, in order to raise funds for the State expenditure. They had paid this assessment for a year without complaining but they had determined to rise, partly to obtain a complete return to the ancient system of revenue, and partly to punish those

> *Kardars*, who had spared neither them nor the Raja, but had plundered both,

Barnes after meeting all concerned and with the consent of the Raja agreed to the wishes of the people for a return to the old system of revenue. A provision was also introduced that, 'if the income proved insufficient for the expenses of the state, the people should make good the differences by a ratable levy, for which the fixed character of Government tribute already assessed upon them afforded a good foundation. The people themselves proposed this rule, anticipating apparently that there would be a deficit'. Thus in 1859 Barnes' settlement came into being providing Rs. 7,809 cash and 18 Karahads, of which the value in money was not worked out. He said, 'no man's burden is grevious although some are much lighter taxed than they should be'. This was the revenue before the settlement of Sham Lal in 1851. Barnes also recommended 'that at a more opportune time the experiment should be renewed'.[13]

Thus in 1874 J.W. MacNabb issued orders for a new settlement, which was carried out under the supervision of the Wazirs, who were to be assisted by six respectable men from six *khunds,* or remote parganas of Athara Bis, Pandra Bis, Bhabha, Rajgoan, Shua and Takpa. This *Waziran* Bandobast remained in force up to the introduction of Tikka Raghunath Singh's settlement of 1894. The net income in this settlement was Rs. 14,255 which included Rs. 12,892 land and revenue Rs.1,363 as cess. Only five *karahads* were retained viz. *mel,* including *soja* or a cess in kind of wheat, rice, *mash bajra,* a cess on all other grains, *pinti*, or a cess in kind of *ghee* and *khora*, a cess in kind of *gur*. The value of five payments in kind thus detailed was in each case calculated and the zamindar had the option of making these payments in kind, or in a fixed cash equivalent. But the Jagirdar gradually imposed all other thirteen *karahads* along with five provided in the settlement on their tenants and thus indulged in gross high-handedness.

Tikka Raghunath Singh's first regular settlement in Kinnaur region was carried out in 1890-1891 by Durga Singh, under his direct supervision. The survey was conducted in accordance with the principles then obtaining in Punjab. A *shajra kishtwar* (field map) and a *misal haquyat* (standing record) were prepared. The settlement report was compiled without dealing with the *Muafis* and preparing the *wazib-ul-araz* (village administration paper). Unfortunately, this report is not traceable now. This regular settlement operation was incomplete in the sense that even the *shojra kistwar* and *misal haquyat* did not fully and duly bear signatures. It appears that some interested parties incited the ignorant people. Rumours were spread that even bee-keeping would be assessed for revenue. The settlement superintendent could not even inspect the revenue estates to judge the assessable capacity. The result of this settlement, however, was on the whole, beneficent. People with foresight saw this though many among the simple masses continued to be restless for sometime. The outcome of the settlement was that the *athwara* (forced labour) was reduced from six months to one month. All the burden some cess commonly called the eighteen *karahads*, were abolished. The land owners granted *adna malkiat* (inferior proprietory rights) while previously even their rights of mortgage and transfer of land were restricted.[14] The net revenue of the region was fixed at Rs. 13,015. A reduction of Rs. 1,240 was made from that of the 1876 settlement. Land revenue was further decreased and fixed at Rs. 10,422 instead of Rs. 12,892 in the previous settlement, but the cess was increased by Rs. 1,230, and was fixed at Rs. 2,593. Thus the burden on the cultivators was decreased and trade was taxed more. The settlement was subject to revision after twenty years. In 1915 when its revision was due, the Raja Padam Singh was not given full powers and the manager Emerson, I.C.S. who perhaps was not prepared to enter into this lengthy business of revenue settlement.

However, Emerson suggested to the Superintendent Simla Hill States for the measurement of *nautors*, i.e. newly acquired land for agriculture. His suggestion was accepted and he completed the work in Rohru Tehsil. But in the meantime he was succeeded by Mitchell who was not in favour of settlement of *nautors* in Kinnaur region as in his opinion such lands were negligible. When the Raja was given full powers in 1917 it was thought undesirable to enter into such a complicated work of settlement. In 1921 the zamindars of Chini tahsil submitted an application for the settlement operations. The Raja immediately ordered it in December 1921. But before it could be actually begun, the people asked for a further postponement. This and several other petitions of the people were accepted by the Raja till 1926.

In 1926 settlement work under the Wazir Chaturbhuj was started. The well-known plain table system of survey by triangulation was adopted in this last settlement, which is still operative. The parganas of Bhabha and Rajgram were surveyed. The Rajgram pargana alone showed 715 *bighas* of encroachments. At this stage there was yet another petition by the people asking for its postponement. On a review of the position, continuance and the completion of the operation was deemed essential if not for nothing else, for the sheer correction of the records. The experience of Rajgram pargana proved that several bighas of land in different parganas had been encroached upon by the people. These *nautors* had to be assessed and entered in the records. Even if the revenue rates were not to be enhanced, a substantial increase in the land revenue was anticipated. In all the parganas all the lands of *blakhal, karali, newal, ghasni* and *bagicha* were assessed and the rate fixed per *bigha* varied from three pies to ten annas. The net revenue of the state rose from Rs. 13,015 to Rs. 21,571, i.e. it increased by Rs. 8,556. Land revenue alone rose from Rs. 10,422 to Rs. 17,257. The cess yielded Rs. 1,721 more.

Government pastures and uncultivated lands were not measured and assessed. Water-mills and oil mills were treated as state property. The rent for *kohlu* was paid to the state at the rate of ten annas eight pies per *kohlu* per year. The rent for water-mills (*gharats* or *ghartangana*) were fixed at the rate of rupees two per year for mills working for 12 months; rupee one per year for mills working for six months in a year; and twelve annas per year for mills working for three months in a year. Similarly, rights of irrigation were duly recorded for each estate. Common lands (*shamlats*) were recorded for common use such as fairs, burial-grounds etc. and were not liable to be partitioned. The Raja was the owner of all forest waste-lands but where deodar and neoza forests lay within the village lands, extensive rights of users were accorded. A nominal tax was levied on trees growing in village waste-lands and new lands could be broken up (*nautor*) without the permission of the Raja. Areas under habitation (*abadi*) were not subject to land revenue and actual possession was considered the extension of ownership of such sites. The ownership for all mines, quarries and streams and rivers vested in the State. Fishing was permitted under licence. Some special treatment was accorded to the border people of Mongoloid origin and Jadakaneits in the settlement of 1894 and the same was allowed in 1928. Keeping in view the difference in cultural pattern etc., the revenue was not settled circle-wise. During 1928 there were only sixty-four revenue *chaks* (estates) but, now due to administrative changes there are seventy-seven revenue estates in the district. Each village has its own deity which is regarded as tutelar god unlike other regions of Bashahr estate where a deity belongs to a *ghori*.

JAGIRDARS AND MUAFIDARS

As elsewhere in the country, in Kinnaur the Raja granted revenue free land to those who were in the state's service, or to the temples located in different villages and these were exempted from *begar* (forced labour). After the formation of Himachal Pradesh certain general principles relating to the administration including resumption of *muafis* and *jagirs* were laid down.

> One of the principle was that the new government would not be legally bound by the grants of revenue made by the previous regime. In the case of grants for service, the primary test prescribed was whether grantee was, and would be any longer rendering any service to the public or the state. In respect of charitable grants, the criterion was whether the grant was and would be, still in public interest. There are at present three grades of *muafidars*. In first grade come permanent ones, in second grade come conditional *muafidars*, and in third grade come assignments to temples. In the first grade there are only three with an area of about 73 hectares assessed to a land revenue of Rs. 127,42 of which Rs. 89.30 represents the amount of assignment. Under conditional *muafidars* there are 70 with an area of about 244 hectares assessed to a land revenue of Rs. 459.37 of which Rs. 141

> are assigned. In third grade there are 104 with 1,243 hectares assessed to a land revenue of Rs. 2,268.30 out of which Rs. 1,580.33 are assigned.[15]

But they were made to pay fairly moderate new cesses from 1948 under the orders of the then Chief Commissioner, Himachal Pradesh.

COLLECTION OF LAND REVENUE

As elsewhere in the country the land revenue in this area is collected by the *patwari*. The rabi collection is made in July and August and in respect of kharif during January and February. All the papers regarding land revenue and the ownership are prepared by the end of October. If necessary he makes corrections in the records by the next harvest, i.e. by June. The patwari's job is to prepare the record of land holdings and revenues paid by each land holder. One of the two such lists prepared by him is given to village headman for collection and the other is kept by him for his record. Patwari's records are subject to checking by the higher revenue officers like Kanungo and Tahsildar. Now regular receipts are issued to land holders by the patwaris of the area. The present land revenue of Kinnaur including seventeen added villages is Rs. 28,654.12. Of this Rs. 23,853.35 is land revenue and the rest is other local rates.

The tenants of temple lands are required to make payment of instruments during worship and in terms of service were required in sowing of fields for the deity, carriage of the deity etc. Generally, rent in kind was half of the produce from which the *Bhoosa* was left to the tenants.

Prior to the enactment of the Himachal Pradesh Abolition of Landed Estates and Land Reforms Act in 1953, several Acts applicable to Punjab were also applied to this region. Under the new Act tenants have secured Bhumidari rights after paying multiples of land revenue. Only ten land owners are affected by the Act and the total area vested in the State Government is only 374.76 hectares of land. A sense of security has come to the tenants and agricultural production has increased considerably. Poor peasants, have thus increased their incomes and are feeling socially well off. The standard of living has increased greatly. With the fixation of the maximum rent at one-fourth of the produce, there has appeared a sense of social justice among the actual tillers of the soil and even some of the landlords have been made alive to the importance and value of manual labour in the form of self-cultivation of land. Wages of the agricultural labourers are now fixed by the state and are subject to revision according to the cost of living prevailing in the region.

Kinnaur like other tribal areas, has been given special concessions in respect of ceiling on land holdings. Here the ceiling on holdings is seventy acres (28 hectares) dry land including orchards, pastures and forests, as compared to ten to thirty acres (4 to 12 hectares) in the rest of the State, depending on the irrigation facilities and number of crops grown in a year. All the landless people were given one acre of land each by the end of September 1975.

Apart from the principal sources of revenue such as land revenue and forests there are other sources that add to the state revenue. Excise is one of the main items. A licence fee of rupees twenty-five is charged from people who distill liquor for their own consumption. Besides local Indian made foreign liquor is available at Kalpa, Puh, Nichar, Sangla and Tapri. This also gives some revenue by way of contract money, Sales, passenger tax, goods tax, court fee stamps etc. are the other sources of revenue of the region.

ADMINISTRATIVE CHANGES AFTER 1948

Tahsildar of Chini tahsil continued to function till the formation of the district in 1960. The district originally had three sub-divisions of Puh, Kalpa and Nachar with three tahsils of Sangla, Morang, and Nachar and two sub-tahsils of Kalpa and Puh. In 1961 Kalpa was made a full tahsil and one sub-tahsil for Hangrang valley was created. Puh was converted into a full tahsil in 1965. The district comprises ninety-five villages in all. The usual resident apparatus of district administration has been planted here with certain modifications to suit administrative and development needs peculiar to the district and different in many respects from certain parts of Himachal Pradesh. It had a single-line administration to ensure rapid development of the Scheduled Tribes till very recently.

Soon after Independence the border region of Kinnaur was administered like that of North East Frontier Agency (NEFA). Here in Himachal Pradesh there was a special unit at Simla under the Secretary to the then Lieutenant Governor who functioned as the ex-officio Secretary (Border) and maintained liaison with the district administration and with various departments at state level and with the central government which was directly concerned with border affairs. After 25 January 1971 when the state was granted a full statehood the above set-up at the state headquarters and the function of various agencies in district became the concern of general administration. Now the separate head of departments correspond directly with their respective district level officers. In general district officers correspond directly with the respective departments but sometimes they route their correspondence through the Deputy Commissioner of the district.

To begin with the Deputy Commissioner was the pivot round whom the entire administration used to revolve. He had wider administrative and financial powers than other Deputy Commissioners in Himachal Pradesh under the system which was known as Single-Line Administration introduced in December 1963. Under this system the Deputy Commissioner had the power to write the annual confidential reports of all the district-level officers. In 1965 the Deputy Commissioner was delegated powers to transfer within the district Class III and IV employees of all departments in consultation with the head of offices concerned. This was primarily to ensure co-ordination of all the departments. The development schemes of various departments have also to route through the Deputy Commissioner when a degree of normalization was attained the district came at par with other districts of the state; except for security matters the schemes of various departments now route through normal channels. Sub-divisional officers and Tahsildars discharge both the functions of judiciary and administration with certain restrictions.

At this point of time there are the following district-level officers in Kinnaur : Police, Medical and Health, Public Works

Department, Panchayat, Forest, Agriculture, Horticulture, Co-operatives, Industries, Education, Employment, Animal Husbandry, Public Relations and Tourism, Welfare, Home Guards, Transport and Treasury.[16]

Several departments of the Government of India are also represented in the district. They are Postal Department, Intelligence Bureau, Area Organizer of the Special Security Bureau, Field Publicity Officer of the Ministry of Information and Broadcasting and Indo-Tibetan Border Police.

REFERENCES

1. From Captain C.P. Kennedy, Assistant Deputy Superintendent, Sikh and Hill States, to the Lieutenant M. Murray, Deputy Superintendent, Sikh and Hill States, dated Subathu, 6 July 1824, *Punjab Government Records: Delhi Residency and Agency, 1807-1858* (Lahore, 1911), Vol. I, pp. 255-99, para 61. Some of the other observations by the travellers on administration are (a) James Baillie Fraser, *Journal of a Tour Through Part of the Snow Range of the Himalayan Mountains and Sources of the River Jamuna and Ganges* (London, 1820), pp. 269-70, 545, (b) Lt. Thomas Hutton, *Journal of a Trip Through Kunawar, Hungrang and Spiti, 1838*, pp. 10, 16; Major Sir William Lloyd and Captain Alexander Gerard, *Narrative of a Journey from Caunpoor to Boorendo Pass in the Himalaya Mountains, etc.* (London, 1840), Vol. II, pp. 301-2, 304.
2. Captain R.B. Nislbet, Superintendent Hill States to Agent to Lt. Governor for cis-Sutlej State, *Foreign Political A*, February 1880, Nos. 406-440. All the documents are from National Archives of India, New Delhi.
3. Col. C. Bedon to Commissioner of Punjab, 11 May 1885, *Foreign Internal A*, January 1886, No. 84.
4. *Foreign Secret, Internal*, October 1919, Nos. 31-4 K.W.
5. *Foreign Political A*, May 1878, No. 145A.
6. *Foreign Internal A*, January 1886, No. 84.
7. Van Der Sleen, *Four Months' Camping in the Himalayas* (London, 1929), pp. 44-45.
8. *Kennedy Report*, 1824.
9. Rahul Sankrityayan, *Kinner Desh* (Allahabad, 1956), p. 313.
10. Himachal Pradesh Panchayat Raj Act, 1968, pp. 56-65.
11. Mamgain, ed., *Kinnaur Gazetteer* (Ambala, 1971), p. 236.
12. From Captain Birch to Lt. Rose, *Foreign Department Secret*, 17 August 1816, No. 5-7.
13. Selections from the Public Correspondence of the Administration for the affairs at Punjab (Lahore, 1868), Vol. 5, Section 4, Memoranda

on the district of Bashahr and the Pacification of the Disaffected Position of its Population CXLIX.

14. Mamgain, ed., n. 11, pp. 241.2.
15. Ibid., pp. 247-8.
16. Administration at district level was discussed in detail with Deputy Commissioner at Kalpa and other concerned officers at state headquarters, Simla:

6

Development and Attitudes

THE AGE OLD isolation of Kinnaur region came to an end with the dawn of freedom in India. The formation of the state of Himachal Pradesh and subsequently the district of Kinnaur on 1 May 1960 was followed by the establishment of several district level offices at Kalpa, the district headquarters. Thus the region became the hub of development activities, which were accelerated by the unexpected Chinese aggression in 1962. Besides, the development of agriculture, horticulture, animal husbandry and cottage industries, communications, education, medical facilities, democratization, community development, power generation are some of the activities of the state in this area. Himachal Pradesh is trying to bring the region in line with the main national life.

COMMUNICATIONS

In the past the means of communication were so difficult that it was well-nigh impossible to enter the region.[1] Perhaps the Kinnaur region was safe from any possible outside attack, and hence the Kinnauras have been able to preserve their distinct culture to this day. The limited means of communication were through the high passes and the valleys of the rivers and streams, primarily the Sutlej Valley. During the first half of the nineteenth century a proposal was mooted to develop a trade road to Tibet through this region for regular trade with Tibet through the path that followed the bed of the Sutlej river or by crossing high passes, some of which are well over 5,500 metres in height. The first road known as Hindustan Tibet Road was developed during 1854 to 1858, and was later abandoned due to the opening of the road to Tibet via Chumbi Valley between Sikkim and Bhutan.

The Hindustan Tibet Road was only a bridle path up to 1960. It enters Kinnaur near Chauhra at Maneoti Dhar beyond Jeori on the National Highway at a distance of 9.6 kilometres from Sarhan (The summer residence of the Rajas of Bashahr) and extends to Wangtu via the villages of Taranda, Paunda, Sungra and Nachar at an average elevation of 2,130 metres. Its length within Kinnaur up to Wangtu is about thirty-four kilometres. From Wangtu it goes up and passes through the villages of Chagaon, Urni (near Tapri) Rogi, Kalpa, Pangi, Rarang, Jangi, Labrang and Kanam up to Shyaso Khad. Its length from Tapri up to Shyaso Khad is about 93 kilometres. It connects most of the villages on the right bank of River Sutlej and is situated at an average elevation of 2,740 metres. From Shyso Khad it passes through the villages of Sangnam, Hango, Puh and near Dabling to Shipki and thereafter enters Tibet. This road remains blocked by winter snows between December and early April. It cannot be used for heavy vehicular traffic; jeeps can ply with great difficulty. In fact it is good for mules and other light transport, such as horses, sheep and goats etc. But the one great virtue of this road is that it connects most of the important villages of the district.

This Hindustan-Tibet Road was the only one mentionable and the rest of the villages were connected by small pathways. Thus the people had to live an isolated existence for centuries away from the mainstream of national life. The economy of Kinnaur was based on simple human labour and unaided by any mechanical devices. Extension of social services was not possible for want of transport facilities and limited comforts could be enjoyed as all that came from the plains on pack animals and human backs, was indeed very costly, beyond the reach of the common people of this region. With the dawn of freedom in India, the development of communication was the first priority for this district. Roads can be the only ready means of communication in this region as railways, waterways and airways are not possible. They are necessary for speeding up the movement of men and material for development activities and also essential for safe-guarding the national borders with Tibet which pass through this region. Now the district has a

National Highway and the several other link roads to the remote regions.

The National Highway No. 22 which connects Simla from Kalka extends to Kinnaur. It enters the district near Chaura Village at a distance of 162.6 kilometres from Simla. It follows the Sutlej Valley first up to Wangtu along the left bank, then up to Karchham along the right bank; up to Shontong along the left bank; up to Dabling along the right bank and thence along the left bank up to Khab where it crosses the Sutlej river to follow the left bank of the Spiti river up to Shyalkhar where again it crosses over to the right bank up to the Sumdo bridge which is the northern boundary of district Kinnaur. Several bridges over the Sutlej are made at Wangtu, Karchham and Shongtong. One each over Pangi Khad, Kashang Khad and Shysaso Khad, are also made. Over Spiti at Shyalkhar and Sumdo bridges for heavy vehicular traffic, are also in existence now. The length of the road within Kinnaur district is 164 kilometres and its width up to Karchham for the first 46.6 kilometres is about 7.3 metres and thereafter up to Sumdo is 6 metres. This is an all-weather road, still in the process of development. The road was aligned by Public Works Department and by early 1961 was made safely jeepable up to Tapri. From Tapri to Sumdo mountain had been cut at various points. With the transfer of the road to the Border Roads Development Board (Deepak Project) early in 1961, the road construction has been carried out by this organization which functions under the Ministry of Transport and the Ministry of Defence, Government of India. This road is now fit for heavy vehicular traffic such as buses and trucks up to its terminus at Sumdo. One great disadvantage of this road for the people is that it passes through only a few villages.[2]

Besides the National Highway No. 22 which passes through the entire length of the district, all other roads are under the supervision of the Public Works Department of Himachal Pradesh. In the district there were 139 kilometres of single lane roads, 200 kilometres of double lane roads, 92 kilometres of jeepable roads, 82 kilometres of tracks and 22 kilometres of metalled roads on 31 March 1979. During the Fifth Plan the Government had proposed to construct 34 kilometres of single lane roads, 6 kilometres of a jeepable road, besides providing cross drainage on the 39 kilometre length. Thirteen kilometres of roads were to be metalled and eleven bridges constructed. The other jeepable roads in the district are as follows :

The Kalpa Line road connects the district headquarters via Peo with the National Highway No. 22 at the kilometre stone showing 226.5 from Simla at Pangi Khad. It is about 20 kilometres long and 4.9 metres wide motorable on which a bus from Simla and another from Tapri runs. It climbs from 1,980 metres to 2,835 metres. It is liable to be blocked for short periods by the winter snows between November and May in its upper reaches, and due to breaches near Telangi village during the rainy season.

The Ribba-Rispa link road was under construction in September 1976 from Kharo suspension bridge over the Sutlej near Khadra Dang.

The Ropa (or Shyaso) Valley Road is 15.85 kilometres long and 3.66 to 4.8 metre wide. It is a jeepable road. It bifurcates from the National Highway at 269 kilometres from Simla and reaches Ropa via Sangnam, Rush Kalang, and Giabond. As it climbs from 2,590 metres to 3,050 metres, it is liable to be blocked by winter snows between December and April.

The Namgya Link road is 5 kilometres long and 2.4 to 3 metres wide. It is jeepable and climbs from the National Highway at a distance of about ten kilometres from Puh. This road provides a link to the last village, Namgia via Khab on the left bank of the Sutlej Valley. Its average elevation is about 2,590 metres and is liable to be blocked by winter snows between December and April.

The Taiti Valley road is 5.6 metres long and 1.8 to 2.7 metres broad. This road enters the Taiti Valley from the old Hindustan-Tibet road about five kilometres beyond Jangi Village to provide a jeepable link to Lippa Village on an easy gradient. It follows the right bank of the Taiti Valley and its average elevation is 2,740 metres. It is liable to get blocked by snows during winter from December to early April.

The Nachar link road is 16 kilometres long and 4.8 metres wide. It branches from the National Highway at 177 kilometres mark from Simla near Sholdang Khad. The five kilometres of this road generally follows the old Hindustan-Tibet road. The latter portion of this road runs at an elevation of 2,134 to 2,438 metres high and is liable to be blocked by snows between December and March. The road follows an easy gradient and a Himachal Pradesh (H.P.) Roadways bus runs over it from Tapri.

The Karchham-Sangla-Chhitkul Road passes through Kamru, Sangla, Rakchham to Chhitkul. This road bifurcates from the National Highway over Karchham bridge on the left bank of the Sutlej and follows the right bank of the Baspa up to the farthest village of Chhitkul. The length of the road is 43 kilometres and it climbs along an easy gradient from Karchham at 1,830 metres to Chhitkul at 3,505 metres elevation. It is expected to be blocked by snows only for a brief period in the winter months from December to early April between Karchham and Sangla, and for longer spells near Rakchham and Chhitkul. H.P. Roadways buses ply on it during the good season.

The Kilba-Shaung Khad road via Karchham is 14.5 kilometres long and 2.1 to 2.7 kilometres wide. This road extends from the Karchham-Sangla-Chhitkul road by crossing the Baspa river near Karchham over the bridge where one branch climbs up the left bank of the Baspa Valley for 9.5 kilometres up to Shaung Khad, while the lower branch extends for five kilometres up to Kalpa along the left bank of the Sutlej.

BRIDLE PATHS

Besides the above network of roads, there are several bridle paths all over the district which connect every village. The notable among them is Rupi Path which connects the village Rupi to the National Highway at Chauhra over to the right bank of the Sutlej river by a suspension bridge. The Wangtu Natpa Path links Kamdar and Natpa to the National Highway at Wangtu. The Rupi-Natpa Path extends from Rupi to Natpa via Bara Khamba, Chhota Khamba, Gharsh and Rokchang-Kachang, over the high mountain to link the villages of Pandrapbis area of Nachar Sub-division on the right bank of the Sutlej Valley. The 16 kilometre Wangtu-Bhabba Path links the Bhabha Village with the National Highway at Wangtu. The Wangtu-Kilba 18 kilometres bridle path extends up the left bank of the Sutlej to Kilba via Ramni Jani and Sholtu. The Shaung-Khad Batsiri 14.5 kilometre forest bridle path extends to Bateri via Sangla. The Shongtong Kamru 26 kilometre path connects the villages of Barang, Mehbar and Kamru to National Highway at Shongtong. It passes over the 3,658 metres high Haran-ghat shoulder. The Shongtung-Tirung Khad 34 kilometre bridle path is under active development for the use of jeeps, and it connects the villages of Purbani, Ribba and Rispa to the National Highway. The Tidong Valley 34 kilometre path from Rispa or Morang to Charang via Thangi and Kuno is fit only for pedestrian-traffic in summer from May to October. The Gyanthing Valley 19 kilometre path connects Nesang village to the National Highway. It crosses Sutlej via wire-rope bridge and then follows the right bank of Gyanthing Valley before crossing the Khad to Nesang village. The Namgya Chango bridle path extends first from Namgia to Tushigang by crossing the Sutlej over a bridge called Rani Bridge. From Tashigang it extends to Chango via Nako and Malling on the left bank of the Spiti Valley. It is open in summer only from May to November, through it the villages of Chango, Nako and Malling are also connected to the National Highway. The Sangnam-Shyalkhar-Sumra 66 kilometre path extends from Sangnam in the Ropa Valley over Hangrang Pass to Hango in the Tirasang-Subsidiary Valley of the main Spiti Valley. From Hango it extends to Shyalkhar via Chuling and Leo up the right bank of the Spiti River and then crosses Lepcha La before dropping to Sumra. It is open in the summer only from June to November. The Taranda-Baruna Khad 80 kilometre forest inspection track extends through the forest belt at an average elevation of 2,740 metres from Taranda to Baura Khad via Paunda, Nachar, Panwai, Rammi, Jaui, Kanai, Sapni on the left bank of the Sutlej and the Baspa Valleys. It is closed from December to May due to snow fall.[3]

This network of roads and bridle paths connects almost every village to the National Highway. As the terrain is difficult and the streams and rivers are many, the traffic across these requires many bridges. Before 1948 there were only a few rickety wooden bridges. The main mode of crossing the river was the *jhula* which consisted of a rope stretched across the stream or river with a cradle running on the rope which was pulled from side to side by guide ropes. With the increasing use of iron the grass rope was replaced by iron cable and nowadays there are bridges over most of the rivers, streams and streamlets. *Jhulas* are also in use at many places where there are no bridges. Over the Baspa river there are bridges at Karchham, Sangla, Batesi and Rakchham. A suspension bridge exists at Rutarang only. On the Sutlej besides on the National Highway, one bridge exists between Chauhra and Rupi and the other at Shaltu for equestrian traffic. Wire rope bridges at Nigulsari, Kilba, Pawari, Akpa, Morang and between Kanam and Nesang exist now for pedestrians only. Most of these rope bridges will be replaced by permanent bridges in due course of time. Some of them are seen under construction. A suspension bridge capable of taking vehicular traffic has been constructed at Kharo and it is half way through at Morang. On the Spiti River two suspension bridges are proposed at Leo and Sumra for vehicular traffic.

Thus the district is now fairly provided with communication. The mode of conveyance is mainly buses and sometimes ponies. Ponies are obtainable only at higher altitudes in Hangrang Valley. All along the National Highway there are several big and small shops which store consumer goods brought by trucks and subsequently supplied to small village shopkeepers. Tea-stalls with a few beds are also available for the travelling public, but they are thoroughly inadequate for the people who use it. No tourist facilities exist as yet except the P.W.D. and Forest rest houses at some places. But now one can drive through the entire length of the district in a single day as against almost three weeks needed fifteen years ago when pony rides were the only alternative to walking.

Post and telegraph facilities were not known to the people of the last century. During the early years of the British regime there were mail runners here also as elsewhere in the plains, for carrying letters from place to place. Earlier descriptions are available in several travel accounts of the Britishers.[4] In 1910 there were only two branch post offices, at Nachar and Chini. Within the next four decades these postal facilities increased gradually during Mr. Emerson and Raja Padam Singh's reign. After the formation of a full-fledged state there has been great expansion of the postal services. Now there is a daily mail service to almost all places in Kinnaur, particularly on the National Highway. Now there are sub-post offices at every tahsil and sub-tahsil headquarters and there are about fifty departmental branches at various places in the district. Though no telegraph office exists as yet, the messages of the people are transmitted by wireless. The first wireless station was opened at Chini (Kalpa) in February 1961. New there is a network of police wireless stations at Kalpa, Puh, Nachar, Leo, Morang, Sangla, Chauhra and Karchham. Besides these the military and Border Security Forces have their separate set for communications with headquarters. A lot more is needed to be done in the matter of communication.

EDUCATION

Until the end of the nineteenth century formal education in Kinnaur was not known. It used to begin,

> . . . by a way of life starting at the mother's breast and the story telling at the grand-mother's knees and not by any institutionalized and literacy-based schooling. This way of life was shaped by challenging natural environments, informed by simple unwritten codes of morality (that form the common foundation of all religions, primitive or philosophical and dialectical as well as all ethical edifices, rustic or sophisticated) and inspired by the example that grand-father set to grand-son, grand-mother to grand-daughter, father to son and mother to daughter.[5]

In some rare cases the head of the family might have taught reading and writing in *Tankari* to his children at his convenience whenever he found leisure hours. Later the Buddhist monks began to teach and preach Buddhist religious education in and around Buddhist temples to such persons as cared to learn, during winter months. This Buddhist education prepared a band of monks and nuns, i. e. Lamas and Zomas as they are locally known in the villages bordering Tibet.

The inspiration for a formal education was provided by the Moravian Mission particularly and the British rule in India, generally. Perhaps the first regular primary school was started at Chini (Kalpa) in 1890 and another at Puh in 1899. Rev. J. T. Bruske and Mrs. Bruske of the Moravian Mission started a school at Puh in 1899, but they left Puh in early 1900. In 1900 they started a school at Chini (Kalpa) for boys and girls which was regularly attended by twelve boys, all very eager to learn. This school at Chini (Kalpa) was upgraded to lower middle in 1920, to middle school in 1944 and to High school in 1952. Next to it a primary school was started at Kalpa in 1914, perhaps by the British manager of the then State of Bashahr. The enlightened Raja of Bashahr Padam Singh who assumed powers in 1917, took a keen interest in education and established many schools in the larger villages of Nachar, Yangpa, Morang, Giabong, Hongo, Kanam, Sangla and Puh. Urdu was adopted as the medium of instruction was prevalent till very recent times with the result that we find Urdu words adopted for the expression of purely local things and customs. Side by side *Tankari* was also taught in some schools. Later on Urdu was replaced by Hindi and *Tankari*, disappeared on its own. Probably the closing of the borders of Tibet by the Chinese contributed to the decay of *Tankari*. Now the medium of instruction is Hindi. Except for the building of the Chini school, most of the schools were housed in private buildings. The problem of getting trained teachers was a major one as no one in those days was prepared to teach in these isolated schools. To overcome this difficulty several teachers were sent to training colleges and normal schools, at state expense. When these trained teachers returned to teach, the position gradually improved. But major problem was the reluctance of the parents to send their children to schools, and hence the enrolment in the schools was very poor. To overcome this problem, the Raja of Bashahr provided stipends and scholarships to poor and deserving students of this region for the pursuit of studies. The aim of the parents in sending their children to schools was only to prepare them for government service. This aim still persists. Students after passing middle standard used to go to the high school at Rampur. At the time of merger of Bashahr state one middle and nine primary schools were functioning in the region.

The period after Independence saw rapid expansion in the field of education. The Education Department, the Welfare Department, the Block Development agency and some voluntary social organizations like Parvatiya Adim Jati Sewak Sangh vied with each other in opening schools and raising the number of these educational institutions still further. Two co-educational *Ashram* schools were established at Rogi (Thakkar Bapa Kinner Ashram) and at Leo (Mahatma Gandhi Ashram, Hangrang). To begin with, the number of students on their rolls were forty and fifty respectively. A similar institution was started in 1960 at Natpa (Pandra Bis Kalyan Ashram) with thirteen students. Two more such schools at Shyaso and Mehbar also came into being later. All these schools were run by voluntary social organizations, but now these have been taken over by the Education Department. At the time of formation of the district in 1960, there were two high schools, four middle schools and sixty-two primary schools. There were twelve high schools, nineteen middle schools and over a hundred primary schools in 1979. In addition to this the State Social Welfare Board is running five Balwadis and five Craft Training Centres. All these institutions are co-educational. The reluctance of parents to send their children to schools has been overcome to some extent by providing stipends and other economic benefits under the tribal and educational funds. The school-level examinations previously conducted by the Panjab University are now conducted by the Himachal Pradesh State Board of School Education, since its inception in 1969-70.[6]

The literacy percentage, according to the 1951 census was only 5.5 of the total population. Out of the total population of 28,972, only 1,509 were literate (1,244 males and 355 females). In 1961 the literacy percentage was 15. Out of the total population of 40,980, only 6,290 were educated (5,503 men and 787 women.) The increase in literacy was encouraging, probably because of the addition of fourteen villages of Simla district to Kinnaur district. In 1971 the literacy percentage of the district was 27.70. In 1981 the percentage is 37.02 according to provisional figure available. Out of the total population of 49,676 only 13,031 were literate (10,819 men and 2,212 females), in 1971, in 1981 out of the total population of 59,154 the number of literates is 21,896 (16,112 men and 5,784 women). It is to be remembered that now there are many outsiders in the district in the government offices and other agencies. The actual percentage of literacy among Kinnauries is still very low. Young children, particularly girls, look after sheep and farms thus keeping the attendance in schools rather low. Some other reasons for low attendance in schools, according to a recent survey in the villages of the Kalpa block are, aversion of parents to sending girls to schools; requirement of children to work at home for collecting wood, for cooking etc.; taking away of children by parents to higher agricultural fields for nearly six months in summer, and the fear that educated children have a tendency to get alienated from the traditional mould and the soil.[7]

In the field of adult education the Parvatiya Adim Jati Sewak Sangh was the first to start night classes soon after the merger of the state of Bashahr. Development blocks also took up the task of adult literacy in 1960 and did very useful work. Adult literacy work was taken over by the Education Department in 1964 and after that the Department opened some adult literacy centres in which several people have been made literate. To provide incentive and inculcate reading habits almost every Panchayat has opened a library of its own, besides a district library at Kalpa. But as all over the country so in Kinnaur, adult illiteracy contributes very largely to the gross illiteracy percentage. Yet, it is hoped that in a decade or so most of the people of the region will be literate and will contribute to the growth of education more actively.

HEALTH

MEDICAL FACILITIES

Difficult terrain with an extreme climate in complete isolation away from the main national life, led people to believe more in the supernatural and depend more upon deity, priest and indigenous medicinal herbs and plants for the treatment of their ailments and diseases. People's faith in charms and incantations as a cure for various ailments was prevalent until very recently. Even now this faith is traceable in the more interior and remote regions. In course of time some private practitioners, trained in Tibet, locally known as *habas*, *chobas*, went from village to village to provide medical relief to the sick. Some of these practitioners were very famous. One of them, Dwarka by name, was known almost everywhere in the region. Dwarka practised the Ayurvedic system of treatment and was a specialist in the use of herbs. His disciples were very famous in the pre-Merger period. The Modern allopathic system of medicine and treatment was introduced by the Moravian Mission at Puh and the Salvation Army personnel at Chini village. Later on, the Forest Department maintained a Sub-Assistant Surgeon with a dispensary at Kilba primarily for its own staff. At the time of the merger of Bashahr state in 1948, only the Civil Dispensary at Chini and an Ayurvedic dispensary at Sangla were in existence. These were only dispensaries and had no indoor treatment facilities. The period between merger and the formation of the district in 1961 has witnessed a significant development of medical facilities by the State Social Welfare Board and their number grew from two to twelve. In 1973 the number of medical and public institutions were more than thirty-six.

Now there is a fullfledged hospital with forty beds at the district headquarters Kalpa, which was only a dispensary opened by the Salvation Army people in Chini village. It has all the important sections of medical, surgical, orthopaedic, ear nose and throat, dental, tuberculosis, pathology, radiology, and pediatrics departments. It has a good staff consisting of a district medical officer assisted by nine medical officers, five compounders and eight nurses. All the serious cases all over the district are now referrred to this hospital for treatment.

Besides this main hospital there are several small and big dispensaries all over the district. Mentionable among them are located at Rupi, Puh, Morang, Kilba, and Tapri. The hospital at Puh is located in the building left by the Moravian Mission and its medical officer occupies the old residence of the mission at present.

Ayurvedic dispensaries are scattered all over the district and their number at present is more than twenty-eight. These are located at Nigulari, Jani, Bhabha, Natpa, Chhota Khamba, Karchham, Purbani, Shaung, Chhitkul, Rakchham, Ribba, Rarang, Giabong, Leo, Chango, Nesang, Charang, Thangi, Jangi, Lippa, Hango, Barang, Sapni, Sumra, Asrang, Nako, Namgya and Urni. Besides these there are several hereditary *Vaids* in some of the villages of the district.

The activities of the Medical Department in the district include the treatment of tuberculosis, leprosy, venereal diseases, apart from maternity and child welfare and the family planning clinics. There are now several lady Health Visitors for the aid and advice of the local population on matters of health and sanitation.

A tuberculosis clinic under a medical officer is located at Kalpa, though the advice and treatment for the disease is provided in almost all centres of treatment in the district. There are four leprosy clinics at Kalpa, Kilba, Nachar and Puh. The special clinics for venereal diseases are at Kalpa, Kilba, and Puh. There are at least five Maternity Service Centres in Kinnaur now. Several Primary Health Centres located all over the district impart family planning advice to the people and its clinics are located at Kalpa, Nachar, Kamru, Kanam and Sangla.[8]

The gradual dissemination of education and induction of outsiders into the district are having a tremendous effect on the outlook of the local people in all matters and spheres. Yet the traditional beliefs and taboos still exist, though they are now losing ground and are being replaced by healthy and modern improved practices.

SOCIAL WELFARE AND BLOCK DEVELOPMENT

Social service organizations and Community Development blocks began their work in this tribal society soon after Independence and particularly after the merger of the state of Bashahr. The prominent voluntary social service organizations are Parvatiya Adim Jati Sewak Sangh, Kasturba Gandhi Smarak Nidhi, Bharat Sewak Samaj and Seemant Pradeshi Society.

The aim of the Parvatiya Adim Jati Sewak Sangh has been the amelioration of the lot of the poor tribal people by adopting various schemes suitable to the peculiar conditions of the area. The Sangh had started Ashram schools which provided boarding and lodging to the students, besides training in crafts like weaving, spinning, tailoring and carpentry. The elementary primary education was also provided to the students admitted and to the day scholars. Now these Ashram schools have been taken over by the Education Department of the State.

Seemant Pradeshi Society began its work after 1974. Its aim is to teach the Bhot language to the Buddhist population of the area. It now runs a school near the village of Chango in Hangrang Valley.

The Khadi and Village Industries Commission has centres at Kalpa, Chango, Rarang, Sangla, Tapri, and Puh for the sale of wool, Khadi goods and other essential commodities manufactured locally at competitive rates. The Commission also distributes spinning wheels, carding brushes, beehives and oil-*ghanis* (presses) at subsidized rates to encourage local artisans in their trade.

The Social Welfare Board of Himachal Pradesh has a Project Implementing Committee at Kapla. There are five Welfare and Extension Centers in the villages of Brelangi, Kothi, Pangi, Rogi, and Kilba. In every centre there is a craft teacher, a midwife and a Gram Sewika. The function of the Gram Sewika is to conduct adult education classes in the evenings, and also to manage Balwadis for the pre-primary children. The craft teacher conducts classes in tailoring, sewing and knitting. The midwife renders assistance in delivery cases and also renders first-aid to the sick in the village. There is provision of a small library for the children of Balwadis and the arrangement for the midday meals are also made. A kind of general education is also given to the people by pointing to the evils of drinking, gambling and other vices.

The wife of the Deputy Commissioner, whenever possible, generally functions as the Chairman of the Project Implementation Committee which coordinates all social welfare activities in the district.

Community Development work through Block Development Agency was launched in 1952 in this area. The Government of India had relaxed the condition of the allotment of the blocks keeping in view the poor means of communications, limited working season, and sparse populations. Usually the blocks were allotted on the basis of one hundred villages with a population of at least 66,000 people. Here in Kinnaur the total number of villages is around ninety-five (only seventy-seven revenue villages) with a population of a little over forty-nine thousand people. Usually a block is put under the charge of a Block Development Officer assisted by a team of experts. The work at the village level is looked after by Gram Sewaks and Gram Sewikas. But in Kinnaur, the Sub-Divisional Officer (Civil) performs the duties of a Block Development Officer in a block with the usual complement of technical as well as village level staff. The sub-blocks are looked after by the local Tahsildars, and here they have not been provided with any extra staff. The main purpose of the Block Development Agency with all its extension officers is to stretch out and unfold the latent tendencies of the people to improve their lot by mutual assistance or self-help. At present there are six Community Development Blocks, and sub-blocks covering the entire district carrying out development activities. Every effort has been made to associate the local people in the work of these official bodies. The more significant institutions into which the non-officials are fully taken into confidence for formulation and implementation of the community development programme include the village panchayats, cooperatives etc.

The enthusiastic participation of the people in extending development programmes is reflected in several ways. They now repair and construct water supply system to their villages. In the field of agriculture the repair and construction work of minor irrigation channels and tanks is done by the people. Their co-operation in repair and construction of bridle paths, jeepable roads, bridges and *jhulas* is remarkable. They also help in the construction of teachers' quarters and nomadic inns. The Block Development Agency also sometimes provides pre-matriculation—stipends, books and stationery, midday meals, housing subsidy, legal aid etc., to the poor people.

POWER

HYDRO-ELECTRIC DEVELOPMENT

The district has a great potential for hydro-electric schemes because of the River Sutlej and its tributaries like Spiti, Baspa and Bhabha rivers. The State Government has prepared elaborate power production schemes and the work has started on some of them. The important hydel projects are the Shongtong-Wangtu Project (400 MW), Topan Powari Project (200 MW), Jangi-Thopan Project (175 MW), and Bhabha Project (150 MW). The Nathapa-Jhakri Project is estimated to cost rupees two hundred and seventy crores. It has already received technical clearance from the Central Government. The Punjab and Himachal Pradesh Governments have also agreed in principle to take it up jointly. This Project alone will generate 1020 MW of electricity. So far forty-four of the seventy-seven revenue villages in the district have been electrified. It is proposed to electrify another twenty-four villages in the current plan period. When this vast hydro-electric potential is fully exploited, the economy of the district will be completely changed. It would then be possible to establish some medium and small-scale industries based on the local timber, fruit and wool resources as a first step towards the development of industries.

POPULAR PARTICIPATION

DEMOCRATIZATION

The autocratic rule of the rajas of Bashahr State was replaced by the democratic set-up with the merger of the State in 1948. Though in the past there were some kind of village panchayats, they were abolished by Tikka Raghunath Singh in the last decade of the last century on the ground of their being partial. Immediately after that people devised the *sianachari* system, on the pattern available in the nearby region of Jaunsar-Bawar, for settling disputes and conflicts. The common belief was that the *sians* were mostly honest, impartial and fair. In this system the aggrieved party used to invite two or three persons, called *sianas*, of the village, to look into his complaint. Their meeting place was generally a place called *santhang*. They, then provided full opportunity to both the parties to put their respective cases and thereafter adjudged the truth or otherwise of the complaint. On reaching a definite conclusion they would pronounce their verdict which was very seldom challenged. This system was in force when the Himachal Pradesh Panchayat Raj Act of 1968 was passed.

Under the Panchayat Raj Act several panchayats were established in 1968. Elections to these bodies were held after every three years. Each gram sabha is empowered to elect on the basis of adult franchise, from among its members, a President, a Vice-President and an executive committee called the Gram Panchayat. Every Gram Sabha is also authorized to elect fifteen adults ordinarily residing within its jurisdiction for the Nyaya Panchayat of that circle. These elected members select from amongst themselves a Sarpanch and a Naib-Sarpanch of the Nyaya Panchayat. The Sarpanch or, in his absence, the Panch approved by him acts on his behalf, for the trial of each case. These Panchayats are given wide powers on the pattern of local self-government, municipal and other similar organizations. In the financial matters they have certain powers to levy taxes, but in practice these are run by the grants-in-aid by the State Government.

Pradhan of a Gram Panchayat is a very important person as he is a link between the people and the State officials. The Government has taken certain steps to make these bodies more effective. For instance, all the Heads of Departments in the government are required to give due weight to the recommendations of the panchayats. Supervision of the foodgrains distribution has been entrusted to the panchayats. Distribution of timber is also made through the panchayats. *Nautor* lands are granted generally on the recommendations of the panchayats are being encouraged to establish panchayat forests and common grazing lands for the benefit of the residents of their areas by giving them necessary land, plants and technical aid.

People's involvement at the grossroots level in democratic functioning has made a marked change in the voting pattern of State Assembly and Parliament. In the general the elections of 1952 for the State Assembly the voting percentage was only

16.82. In the election of Territorial Council in 1957, the percentage rose to 25. In the third general election in 1962, the percentage of votes polled of the total number of votes was 37.69. In the fourth general election of 1967 the voting percentage rose to 67. This was really creditable for such a mountainous and backward area. Thus the people's involvement in the democratic functioning has gone into the way of life of the inhabitants of Kinnaur.

The massive development work done by the state and voluntary social service organizations have opened vast avenues for the people. Besides government service, from the competitive all-India services down to class four cadre of employment which the tribal young men may now try to enter, chances of semi-official and non-official employment have increased many times. The various development programmes have brought along a very large increase in the population because of the influx of workers from the neighbouring districts and because of which further governmental spending is needed for developmental, administrative and security purposes. As a whole sources of commercial earning, from petty tea-stalls to impressive shops, and other ventures have shown an upward trend which is expected to continue in the future. The people on the whole feed, clothe, house, educate and medically treat themselves much better and communicate farther and much faster and much more than before.

Kinnauris, initially reluctant to accept change in their age-old values of life, which had to a great extent been instrumental in keeping them backward, are now busy building up a new life for themselves and new pattern for society, preserving the rich cultural heritage at the same time. Signs of diversified rural economy and richer and fuller life are already visible. It is hoped that in no distant future the region will be in step with the national development in all walks of life. However, the problems of increase in crimes and alienation of the younger people from the land which the older people have started complaining about and unemployment among the educated young men would take some more time before they are solved.

REFERENCES

1. The conditions of communication in the early 19th century are given by Alexander Gerard in his book *An Account of Koonawar in the Himalayas etc. etc.* (London, 1841), pp. 36-7.
2. On National Highway No. 22. There are boards displaying the distance of places from Simla and their level. One such board is located at Pawari petrol pump.
3. All information regarding roads and bridle paths under P.W.D. is available with the Executive Engineer, Kalpa. When I visited the region Mr. Handa an inhabitant of Mandi was posted there.
4. Dr. W. G. N. Van Der Sleen. *Four Months Camping in Himalayas* (London, 1929), pp. 57-8.
5. T. S. Negi, *Scheduled Tribes of Himachal Pradesh : A Profile* (Simla, 1976), pp. 50-1.
6. Information regarding education are collected by an old teacher Mr. Ram Saran Negi of Telangi village near Kalpa and several Headmasters of Middle and High Schools located in the district, besides officials of Education Department at State headquarters.
7. Literacy figures and other data is drawn from the Census Department publication.
8. Information regarding medical facilities are collected from the District Medical Officer, Kalpa, a resident doctor at Puh and a Vaidya at the village Chango. Aid from government officials at Simla has also been taken at times.

7

Reflections and Conclusions

KINNER DESH of mythological fame is the modern district of Kinnaur. It is located on the international boundary between India and Tibet-region of China, west of Nepal. The district is located between the districts of Lahaul Spiti and Uttarkashi. In the north-east there is Tibet and in the south-west lies the Simla district of Himachal Pradesh. The region represents all the three main high ranges of Zaskar, Great Himalayas and Dhaula Dhar.

This region, popularly known as 'Sat Khand', consists of seven valleys in its fold. Sutlej is the main valley which runs all along the district dividing it into almost two separate parts. Besides Sutlej, the valleys are Spiti, Ropa, Bhabha or Wangpo, Tidong and Baspa. Baspa is the most beautiful valley and in contrast Spiti is the most barren. In between the two are located the remaining five valleys.

This whole region is watered by three main rivers, the Sutlej, Spiti and Baspa. These are fed by eight small rivers : Yulang, Ropa, Pejur, Kashang, Mulgaon, Yula, Wanger and Tirung. The climate is mostly cold except for a very short period that may be called summer by the locals between June and September. The area is both wet and arid. South of Kalpa, the region receives rainfall but beyond that it is dry. This contrast is clearly visible from Spilo village on the National Highway No. 22 which runs via this district. The average elevation of the region is between 1,220 and 6,770 metres above sea-level.

This rugged land-mass with uneven terrain is spread in 6,553 square kilometres, with a total population of 59,154 heads. The population per square kilometre is only nine. Administratively it is divided into six tahsils, viz. Sangla, Nachar, Kalpa, Morang, Puh and Hangrang. The district headquarters is Kalpa, located in the shadow of the famous Kinner Kailash.

The inhabitants of this region are different from the Khashias of western Himalayas. The Kanaura or Kinnara tribe are the original inhibitants of this region. The people are well built with fine Persian features in the region south of Kalpa, and in the north the features of the people are slightly Mongoloid or, more appropriately, Bhot. The ladies of Sangla and Kalpa regions are so beautiful that their parallel is difficult to locate anywhere in India. They are appropriately known to have sweet voices—*Kinner Kanthi*. Men are of a normal size between five feet two inches to five feet ten inches. They are simple, honest, hardworking, enterprising and hospitable. These qualities have been praised by all those who have come in contact with them from the earliest known times.

The usual caste structure as obtains in other parts of India is found here also. Kanets or Negis are the upper castes and the lower castes consist of Chamang, Domand and Oris. Equivalent to Rajput, Lohar, Badhai or Tarkhan and Ko.is, i.e. Rajput, blacksmith, carpenter, and weavers. The main religions followed by people are Hinduism and Buddhism. The villagers in Hangrang valley bordering Tibet and a few in Puh tahsil follow Buddhism. Most people in these villages have two names, one Buddhist and the other Hindu, given them by lamas and the other by mostly teachers in the schools. There have never been any brahmins in this area. Hence religious-ceremonies are performed either by lamas or negis (Kanets).

There have been several Christian missions in the region during the last century and in the present one, but the absence of the Christians is conspicuous. It is difficult to explain the precise reasons for their absence, but one can hazard a guess. The region from time immemorial followed the established religions of Hinduism and Buddhism. Hence it was difficult for any outsider to make any dent in the beliefs of the people. Secondly the region was part of the native state of Bashahr whose Raja was a Hindu and a follower of the brahminical tradition. People generally believed in the divine origin of kingship and hence they mostly followed their Raja in the matter of religion, except for the regions which are near the Tibetan border. Thirdly state patronages to the obedient might have inclined people towards the established religion followed by the Raja. Finally the state might have secretly warned and prohibited the people to adopt religions other than that of the Rajas. After Independence and particularly after the Chinese occupation of Tibet the situation has dramatically changed. The Buddhist patron of Potala is no longer there. The border has been sealed and the Dalai Lama is a refugee in India. This has made some impact on people's religious beliefs. Now more and more people are tempted to believe in Hinduism. It seems, there is a positive shift, in the religious beliefs of the people and the focus now has come round to Hinduism perhaps even in the border villages of the region.

People by and large speak the Kinnauri dialect, called Homskad, but now educated people, particularly below the age of twenty-five can speak Hindi; a few among them can speak English also. Tankari script of the bygone days is no longer in use. It is possible that it is used now for religious purposes in the monasteries. People live in small villages located on the slopes of the mountains. The houses are of two storeys generally. The lower apartment is used for keeping the cattle. These houses have only one living room with balcony attached, but now the new constructions have more than one room. These can be seen at Kalpa, Koti, Duni and Chango villages. Except for a few household things most of the houses do not have cots to sleep on. People by and large sleep on the floor on *khera* or *Kharech* and *Pekpa*, i.e. mats woven of goats' hair and skins.

Due to the cold climate of the area people wear woollen clothes throughout the year. The *thepang* or a woollen cap with or without *chambaka* flowers is the common head wear which is used by women and men alike. The men wear woollen shirts *chamu-kurti*, a long coat (*chhuba*), woollen pyjamas (*chamu sutan*) mostly of grey colour. Ladies wear woollen *sari dhori* full sleeved blouse with a *choli*, *chanli* or *shawl*. This shawl is wrapped round the shoulders and its two ends are fastened together near the breast by means of a silver hook called *digra*. Women often wrap round their waist a scarlet coloured woollen or cotton cloth of about five to eight metres in length and about a metre in width. This is called *gachhang*. On the occasions of marriages, fairs and festivals, people put on embroidered clothes in various designs and colours. The ornaments generally used by the women are : *mul-u*, *taamani*, *kantai*, *khandoch*, *chalin*, *shanglayang*, *digra*, *tomukch*, *laung* and *daglo* mostly made of silver. Except for the *murki* in the ear and rings on the fingers the men do not use ornaments in this region.

The opening up of the area due to increased communication channels has made a positive impact on the people. Now more and more men are using modern suits, coats, pants, shirts and other clothes used in the plains. The ladies have taken to wearing salwar and kurta, but the use of sarees is not seen yet. In place of the old *pona*, i.e. local made shoe, people now are using shoes made either in plains or by the local manufacturers. In place of cow-dung wash, some houses are now painted with modern paints. Radios, transistors, watches, clocks and other luxury items are now available in the area. P.W.D. guest houses are equipped with heaters and geysers.

The people in this region have their meals three times a day. The staple food is wheat, *ogla*, *phapra*, barley and pulses like peas, black peas, beans, *mash* and *masur*. Besides these, people use vegetables which are available, like turnip, cabbage, peas, pumpkin, potato and tomato. Kinnauras relish meat; goat and ram meat is mostly in use. Chicken is taboo in the area. Drinking of liquor is quite common in the district. In fact every household has a permit for the production of liquor (*ghanti*) for domestic consumption.

Now most of the food stuffs are readily available in the region; either fresh or tinned. This has had some impact on the food habits of the people. But this impact is so marginal that it cannot be called a change. However, among the younger generation, particularly among boys up to twenty, there has developed a craze for new things. Prohibition has also made some marginal impact on the educated yong men.

Cinema and other urban recreational facilities have not reached this region as yet. Hence people depend mostly on the traditional means of entertainment. These include music, dancing, fairs and festivals. All round the year, almost every month some fair or festival is celebrated. During winter young unmarried girls celebrate for a week or more, a festival known as *Toshing*. At this time they invite their boy friends over a feast prepared of special food and drink, and eaten it together for a week or so. There is a great deal of singing and dancing. These dances at times continue till the early hours of the day.

People all over the world believe in superstitions, and dreams and have faith in the supernatural etc. Perhaps in this case, because of this difficult, inaccessible terrain, people are more ready to accept them than the people in the plains. The form content, and style of such beliefs may differ from region to region, from country to country, but they continue even in the present age of scientific development.

The moral values of the people here have been a controversial subject. It has invited several adverse comments by travellers who happened to visit the area in earlier times. Until recently monogamy was the exception rather than the rule in the region. Polygamy is not unknown in the area. Polyandry however has been the rule in the most parts of the region. In this system several brothers share one wife. In case the joint wife proved to be barren, her sister was brought in as a second wife. Sometimes the younger brother might prefer to bring in another wife for himself on the plea that the common wife was too old for him.

In the changed circumstances of today the institution of polyandry has come under severe strain. It worked fairly well when all the brothers in a family were engaged in different vocations and when they used to live away from home taking turns so that one alone remained at home. Now that trade with Tibet operates no more, cattle feeding is done virtually at the doorstep, and P.W.D. work is readily available in the village, all the brothers in a family remain at home and they often quarrel among themselves making the life of the common wife quite miserable. Hence the institution of polyandry is on its way out. People are leaning more and more towards monogamy. This might, in due course, affect the land holdings and the very structure of the family. Already small land holdings may become smaller.

Rituals observed in this region are mostly the same as we see in the plains. Ceremonies at birth and afterwards like hair-shaving (*mundan*), *krachogmig* are all performed with the same religious and social fervour as it is done in the plains. However, in the border regions there are two ceremonies that are different and peculiar. These are *khaskis* or *bedhai* and *bose*. In *bedhai* all the clan people come and congratulate the parents with some presents. The *bose* function is held between one and three years of child birth. In this local deity is invited to bestow blessings. On this occasion also a big feast is arranged by the parents and all those present participate in the dances and music on that day. In all other regions *Gangajal* (water from Ganga) and *gomutra*, is used or *hawan* is performed for purification of a place or person.

Death normally takes place on the ground as people generally do not use cots. But in case a dying person is on a cot or bed, he is brought down on the floor before death. Just before death *panchratan* (an alloy of gold, silver, copper, brass and iron) is put in the mouth of the dying one. After death the body is taken in procession and cremated. Ceremonies after death are performed as it is done in Hindu families of the plains.

There are four forms of marriages in the region. The standard marriage is known as *janekang*. It is considered as wedding. In this form marriage is properly arranged through the middle man. The other form of marriage is *dimtangohis* or *bennang-hachis* or *jushis*, i.e. love marriage. In this, boy and girl decide to get married and the boy brings the girl to his house secretly. The third form of marriage is known as *dubdub* or *khuchis*. In this form boy either alone or with the help of friends captures the girl either from fair or from farm and carries her to his house. In both these, i.e. in love and capture, later a standard form of ceremony is observed. The fourth type of marriage is known as *Har*. In this form a married lady is taken away by other man after making the payment of expenditure incurred in the marriage by the former man.

Dowry is unknown in the region, but presents are given to the bride by her parents. From the groom's side those present give some money to the groom's father. This money is used by the groom's father. This custom is called *baildavang*. Similarly, every one present from the bride's side gives some money to the bride. This collection is the exclusive property of the bride and the ceremony is called *udanang*.

There is a unique custom in the tribal society of Kinnaur that after the marriage the father-in-law gives some land and a room in the house to the bride. This is written on paper in the presence of a number of people and this paper is kept with the father of the bride. In case of a dispute with the husband the bride gets that part of house and land. This is known as *bandobust*.

II

People generally believe in the tradition current in the region that ancient Kinner Desh was a place where the five Pandava brothers and Draupadi, their wife in common lived incognito during their exile. Some object or the other that appears ancient is associated with them. It may be a wooden gate or a temple or a tank or a gate on the roadside. People say that the Pandavas used to work in the night and during the day time they went into hiding, in the caves or in the deep valleys. If the tradition has any basis in truth, it is not very clear whether the inaccessibility of the region or the general hospitality of the people had tempted them to live here.

However, it is certain that the Kinner people have deep roots in Indian mythology, legend, and literature. They are generally considered as a distinct race. We find mention of these people in almost every ancient text, be it the *Vedas* and *Puranas* or legal work of Manu, *Manusmriti* or literary work of Kalidasa, *Raghu Vamsha*. They are mentioned in both the *Ramayana* and the *Mahabharata*. In some texts they are called Kinner while in others they are known as Kimpurusha. Some call them semi-men and semi-gods, some others call them half men and half animal. These contradictions might have crept in from the *Jataka tales*. These people are remembered as happy people enjoying dance and music, both vocal and instrumental. The legendary, mythological, and literary accounts of Kinneras may to some extent be true, but even if they are not, they most certainly indicate the very highly developed creative and inventive genius of people in general and the authors of these texts in particular.

Neither Kinneras nor Kinner Desh is mentioned anywhere in the recorded history. The area presently known as Kinnaur district must have been part of the mighty empires of Mauryas, Guptas, and Harsha, till about the middle of the seventh century A.D. It is only after the fall of the dynasty of Thaneshwar and Kanauj that we find the growth of small principalities. As elsewhere in India, similarly on the border regions, there grew small states like Gilgit, Baltistan, Ladakh, Spiti, Bashahr, Garhwal and Kumaun. Kinnaur was a part of the former state of Bashahr till the dissolution of the state in the forties of the present century. The genealogy of Rajas of Kamru and Bashahr indicates the establishment of the state in the sixth century. Who these rulers were, it is difficult to say, but it seems certain that they were the emigrants from the middle country.

It is mentioned by Rahul Sankrityayan that the region was part of the Bhot empire during the seventh and tenth centuries. But from the close examination of events and the nature of Tibetan imperialism, it appears that their thrust was upwards all along the river Indus, leading them to Yarkand and Khotan in Central Asia. Had there been any crossing of the River Sutlej by Tibetans ? Then what kept them from reaching the fertile plains of India, which were then dotted with petty principalities ? How is it that the people of Kinnaur have been able to preserve their traditional culture to this day, if there had been Bhot rule for so many centuries as mentioned by Rahul Sankrityayan ? To me it appears that the fiction of Bhot rule in Kinnaur is thoroughly misleading and erroneous ; unfortunately this fallacy has found place in the Gazetteer of Kinnaur and in other popular works on Himachal Pradesh.

During the medieval period, some of the hill states such as Kangra, Chamba, and Sirmaur were attacked and made tributaries to the Mughal emperor at Delhi, but the state of Bashahr could not be reached by any adventurer of that period, Raja Kehri Singh was by far the most popular and powerful ruler of the state. He had not only consolidated the state by adding several regions to it, but had also participated in the

Tibetan-Ladakhi-Mughal War of 1681-1683. In this war he sided with Tibet and at the end of the war he received the portion of the territory beyond river Sutlej which was occupied by them during war, as reward for his help. In a commercial treaty with Tibet he obtained free trade and movement of Bashahr people in Tibet without restriction. Raja Kehri Singh was granted the title of 'Chhtrapati' and a Khillat (Robe of honour) from the Moghul Emperor Akbar.

Subsequent to the death of Raja Kehri Singh, Bashahr involved itself in the petty quarrels of the region. The small disputes continued till the rise of Gurkha power under Prithvi Narayan Shah, the Sikh power under the celebrated Ranjit Singh and the British power under Richard Colley Wellesley. In 1810 the state of Bashahr was occupied by the Gorkhas and the Raja had to flee to Kinnaur region of the state. He remained there till after the settlement of war between Nepal and the British East India Company in 1815.

Raja Mahendra Singh was restored to power by the British Government and was granted a *sanad* in November 1815. He was a minor at that time, therefore the administration was carried on by the hereditary Wazirs of the state. On becoming a major, he proved to be a very weak ruler. His son and successor Raja Shamsher Singh was no better. Hence the rule of Wazirs continued till 1886 when Tikka Raghunath Singh took over the administration of the state. Till his death in 1898 the state was well managed. After his death the British Government took the administration in their hands and a manager was appointed to look after the state. After three years of the succession of Raja Padam Singh in 1915 the rule of the manager came to an end. Soon after Independence, the state was abolished and the Raja was pensioned off. Padam Singh died soon and was succeeded by Vir Bhadra Singh, who was deputy minister in Mrs. Gandhi's Government member of Parliament.

III

The people of Kinnaur had to work hard for their livelihood. Agriculture, animal husbandry, horticulture, forestry, cottage industry, trade and commerce must have been equally important for them, from the earliest known times. This fact had been noticed by those who went to this region. Even now the people have to work hard for livelihood.

Agricultural holdings are very small. The area is cold and hence the timetable of crops also differs from the plains. *Rabi* crops are usually sown from the middle of September to mid December and reaped between April and July, depending upon the altitude of the place. Similarly, *kharif* crops are sown from March to middle of July and reaped from September to the end of November. The chief *rabi* crops are barley, wheat, lentils, coriander seed and field peas. Other smaller crops are *phapur*, *bathoo*, *cheema* and *kodu*. *Kharif* crops are millets, potato, and maize. *Chicke*, *chit* and *chid* are sown in July and cut in September and are used usually like rice. The area has been and still is deficient in food production. In earlier times supplementary diets like dried apricots and *chulis* were used. But now due to the opening up of communications, the products of the plains are readily available there. Besides, Government has taken several measures to improve the production. Seeds, fertilizers, and *takavi* loans are now readily available to the farmers.

Animal husbandry is one of the main pillars of the tribal economy. Its importance can be judged from the fact that in 1972 the human population of Kinnaur was 49,835 and the gross livestock population was 1,25,708. It comes to 2,552 per thousand population, out of which 1,297 were sheep, 669 goats and 444 cattle. Chief among the domestic animals are horse, mule, donkey, yak, cow, ox, sheep, goat, mare, dog and cat. The major occupation of the people has been sheep and goat rearing. The benefits derived from these sheep are manifold. Besides providing milk, meat, skin, and wool, these animals were used as pack-animals from very early times. With fast development of means of communication their use as pack-animals is very much restricted. These animals are the most valuable assets of the people of Kinnaur. Poultry farming was unknown to the people of this area, as eating chicken was taboo. Now some farms have been started at Peo and Kalpa. Government has taken several steps to improve the livestock. Veterinary hospitals have been opened and cross breeding with

Spanish and Russian merinos was taken up some years ago. In order to protect sheep from skin diseases, sheep-dip centres have been opened at important routes.

Horticulture is yet another source of income to the people of the area. The land is ideal for the growth of *chilgoza* trees which have been nurtured there from time immemorial. Fruits like apple, apricot, peach, and chestnuts were also available. Now several fruit research stations, demonstration orchards, and nurseries have been established to develop horticulture in this region. The annual yield of apples, grapes, chestnuts and walnuts have increased considerably. It seems what is needed now is to provide quick transport and packing materials at convenient centres. It is a common sight to see people bring their apples to the roadside in the season and sell them to the local shopkeepers for a mere twenty paise per kilogram.

Forests are yet another source of income. In fact the total population of this district depends entirely on the forests for their day to day needs of firewood and timber. Besides valuable *chilgoza* trees which fetch a handsome price, these forests produce several medicinal herbs such as *kark*, *ban*, *kakri*, *kalazira*, *belladona*, *dhoop*, *chirata*, *cirtemisia*, *ephedara* etc. A concerted effort is needed on the part of the State Government to explore the possibilities of cultivation of herbs on a commercial scale.

Dhori, *namda*, *gudma*, *pattu* and several other woollen items are manufactured here by the people. In fact all local needs of cloth are met by the local produce only. Nowadays artistic handicrafts produced by artisans individually or in the cottage industry have a ready market at Simla and other centres of handicraft sale. There is no dearth of natural resources in the area. In fact deposits of iron, silver and other minerals were discovered before the formation of Himachal Pradesh. What is needed is their proper use on a regular basis. Bee-keeping, and manufacture of liquor for local use are some of the many activities of this area.

Many production-cum-training centres have been established for various crafts by the Government. Incentive is now provided by granting loans to the people. Articles produced in cottage industry are kept for sale at Kalpa and Simla in the emporia.

Trade with Tibet was one of the most important sources of income to the people. In the bygone days when money circulation was very limited the mode of exchange was by barter. In course of time a limited currency was introduced. Traders generally used to purchase goods from the Lavy fair at Rampur and get them exchanged at Garo and Leh by the products of Tibet, Yarkand, and Ladakh.

To begin with, the local deity was their bank. In course of time the needs of the people increased for various domestic purposes. Large landholders and local merchants (*sahukar*) began to advance money to the people. With the dawn of freedom the big landed estates were abolished and hence the local zamindar could not continue lending money. When border trade was stopped due to sealing of the borders by China in 1958, not only people but cattle also suffered. But soon after, local grazing grounds in Himachal Pradesh were shared by people. Traders started small shops on the roadside and in the villages. Large-scale constructions of roads and other office buildings in the area provided work to the people at their doorsteps. Some big traders formed co-operative societies and started supplying to the military establishments in the area. Banks and co-operative societies have also reached the area. Fruits and other local products are now sold readily which generate capital in the region, and that in turn is used by people for improving their living standard.

IV

How ancient Kinner Desh was administered, is not known. The nomadic pastoral people of the region must have gradually settled in the small villages with their deities. In course of time, the most powerful amongst them must have grown to the position of Raja or king. Amongst kings one might have emerged more powerful than the rest. This king could be Banasur as tradition tells us. An autocratic polity must have emerged in due course. But it seems that a certain amount of independence was available to people who used to organize themselves in the villages. Some kind of panchayat system was available in the villages. In the beginning there were three tahsils in the state of Bashahr. Each *tahsil* was divided into *parganas*, *ghoris* and *villages*. In Chini Tahsil there were 5 parganas, 14 ghoris and 63 villages. Now the number of villages is 77. This whole area was under a Wazir known as *Sarhadi* Wazir. This system was continued for a long time in the state. Later for a brief period of about eleven years, Tikka Raghunath Singh was the real ruler of the state. After his death a manager was appointed and the manager's rule continued till 1917 when Raja Padam Singh took over the state administration. After the dissolution of state, Kinnaur became a part of Mahasu district. Early in 1960 a separate district of Kinnaur was formed and now usual district administration is obtainable in the area.

Crimes were never a problem in the region. Even now there is only one small detention cell at Kalpa, which remains empty almost all the time. Crimes involving women are of course there, but they are mostly settled amicably. Now cases of minor thefts are also reported. A full-fledged judicial machinery from *Nyaya* Panchayats to the court of district magistrate is now available for dispensation of the justice. Cases from district courts go to the High Court, situated at Simla.

Revenue was realized in kind by the Rajas of the state. Anything and everything which was good was appropriated by the Raja. *Begar* was legal. But in due course of time, the demand for cash revenue was made by the people. Therefore the first revenue settlement was made by Sham Lal in 1854. This settlement became a bone of contention between Raja and his Wazirs. Sometimes revenue was realized in kind and sometimes in cash. Cash revenue was beneficial to Raja whereas revenue in kind was in favour of the Wazirs. This tussle resulted in several Revenue settlements. These were Lord W. Hay's settlement in 1856, Mr. Barnes's Settlement of 1859, Mr. J.W. MacNabb's settlement of 1874, and Tikka Raghunath Singh's settlement of 1890-91. Tikka's settlement was somewhat permanent and was in force till 1928 when a revised settlement was done. This settlement of 1928 was continued when the state was abolished and the Himachal Pradesh was formed.

With the formation of the district, the people have come in direct contact with the authorities who are responsible for the administration. They are now in a position to petition and obtain justice. They are now made aware of the several beneficial schemes of the government, and they derive benefits from them. Be it a Takavi Loan for development of agriculture or a co-operative loan for other purposes it is readily available to the people.

The old isolation of the region is now broken by a network of road communications in the region. Besides a National Highway which connects Kalka to Kaurik a distance of about 441 kilometres, which runs through the district, there are several feeder roads and bridle paths which connect almost every village to the national highway. There is no doubt that the villages situated at the old Hindustan-Tibet Road are now losing their importance. The new centres are coming up and markets like Chango, Puh, Pawari, Karchham, Tapri and Chegaon are gaining importance by virtue of their situation on the National Highway No. 22.

Education has received greater attention after Independence. In place of few schools, there are now almost as many schools as the villages. Besides this there is a High School at Kalpa and several middle schools. Though no college is there, but there is a great need of one at Kalpa. I have come across several boys who had passed out of High School but could not proceed for higher education as there are no colleges in the area.

Medical facilities are now available in most of the villages. Besides Vaidyas, primary health centres, maternity centres and dispensaries are now available to the people. At the apex a district hospital with modern equipment is located at Kalpa, the district headquarters.

There are several social welfare units which are active in the region. Prominent among them are Semant Pradeshi Society, Khadi and Village Industries Commission, Social Welfare Board, and Community Development Block agencies. People's co-operation is readily available for all these units and hence they have established several welfare centres in the region.

There are now several hydro-electric stations in the region. Prominent among them are Shongtong-Wangthu Project; Thopan-Pawari Project, Jangi-Thopan Project and Bhaba Project. The Nathpa-Jhakri Project is estimated to cost rupees two hundred and seventy crores. Out of the seventy-seven villages about forty-four are electrified by now. It is hoped that every village will get power in due course. When this vast hydro-electric power potential is fully exploited, the economy of the district will be completely changed.

V

The changes brought about by the formation of the district and the opening of communication are many and far-reaching. In the family structure there is a major change. Polyandry is gradually on its way out giving way to monogamy. In the matter of inheritance, in place of right of primogeniture, equality of the rights is being established gradually. Superstitions faiths and beliefs are gradually giving way to reason and rationality. Though until now fairs and festivals are the only forms of entertainment, in due course of time cinema and theatre will go to the region and perhaps people will have more recreational facilities than they do now.

There is no doubt that people are now better fed and clothed than they used to be though a lot more remains to be done. Natural resources are yet to be exploited fully. Communications are to be developed in such a fashion so as to reach every village. Only then will the local produce of cottage industry and horticulture reach the market speedily. Agriculture can get a big boost only if somehow the water from the rivers can be lifted to the upper regions. It is hoped that in due course, with the development of the economy, lift pumps will be installed at convenient points at almost every river, particularly Spiti River which runs through the completely arid zone of the district.

With the advance of education and democratization it is expected that understanding will develop more rapidly than before. More and more people will participate in the process of development of their region and the surroundings. Boys will find more places in competitive service examinations and other technical and utilitarian jobs of the state and the country. The time is not far off when these tribal people will join the national mainstream in all matters as equal partners.

Appendices

APPENDIX I

HOUSEHOLD UTENSILS

Local name	English name
Korchhi	Ladle
Lamthu	A big brass or copper vessel
Bodu	A round shaped cooking vessel
Digch	A little kettle, a small pot
Dig	A kettle, a pot
Lothi	Cauldron
Tusna	Iron hook
Chharpona	Sieve
Ron Pan	Iron griddle
Parat	A big iron pan
Kunal	A big brass pan
Chokthal or *Tasla*	Basin
Musarbo	A jar for wine
Kargyul	Silver cup
Batich	Cup
Nangch	Small dish
Nang	Platter
Galas	Tumbler
Lotri	Pot, lotah
Surahi	Pitcher, jug
Gagri	A vessel for containing oil etc.
Chimto	Tongs
Arbo	A vessel for washing hands
Ladanang	Iron stripod

Handia	Earthen pot
Gor	Earthen vessel
Dongbo	Tea Churner
Kaning	Mortar
Challang	Sieve
Ponach	A small ladle
Khyot	Big spoon
Dakna	Lid
Thumbu	A big brass ladle
Zom	A wooden vessel for carrying water
Chadig	Tea pot

APPENDIX II

ORNAMENTS

Name of Ornaments	Part of the body where used	Metal
Chak	head	silver
Tanol	head	silver
Zuti	head	silver
Pipla	head	silver
Tunki	head	silver
Tanole	head	silver
Dabmige	head	silver
Mul-u	ears	silver
Kantai	ears	silver or gold
Jumku	ears	silver
Khul-Kantai	ears	silver
Kante	ears	gold
Laung	nose	gold
Balu	nose	gold
Khundoch	nose	gold
Balak	nose	gold
Likcha and *shulikcha*	neck	coral or some precious stone
Patkachang	neck	silver
Chandra-malang or *Chandra-har*	neck	a silver garland containing beads of coral turquoise etc. and old silver coins of four annas, eight annas and a rupee
Tungma, gou-tungma and *chostung*	neck	front in silver and back in copper
Taamani	neck	necklace having three large gold beads
Kontmala	neck	silver
Daglo or *taro* or *brindlo*	wrist	silver
Kagun	wrist	silver or gold
Bang-pole	toes	silver

Digra	breast	silver (to fasten the two ends of shawl)
Tomukch	breast	silver (to fasten the two ends of a *dhori* near left shoulder)

The Ornaments of Upper Kinnaur

Bairak	head	It is made of cloth, *mul* (silver), *yu*, *tungrek* (white stone) *ragan* (brass), *churu*, *poshal* (yellow plastic and *sar* (gold)
Konta	ears	silver and (coral) *moonga*
Menthak	ears	silver
Lonk or *laung*	nose	gold
Tirmol	neck	silver and *moonga* (coral)
Gaun	neck	silver, gold and *pheroza* (turquoise)
Ulik and *Konthi*	neck	*moonga*
Pashal or *Gyamrang*	Fastened on shoulders with the help of hooks	plastic, (coral) *moonga*, copper and silver
Dugu	wrist	silver
Dunglak	wrist	broken shell
Surtuk	finger	silver
Digra	breast	silver brooch

APPENDIX III

KINSHIP TERMS

Relationship	Local Terms
Father's father's father	*Kotete*
Father's father's mother	*Koapi*
Father's father	*Tete*
Father's mother	*Aie*
Father's elder brother	*Tegbaba*
Father's younger brother	*Bapuch*
Father's elder sister	*Nane*
Father's younger sister	*Nane*
Father's elder brother's wife	*Teama*
Father's younger brother's wife	*Amnich*
Father's elder sister's husband	*Mama*
Father's younger sister's husband	*Mama*
Father	*Boa, Boba*
Mother	*Ama*
Mother's father	*Mapo-tete*
Mother's mother	*Mapo-Aie*
Mother's elder brother	*Moma*
Mother's younger brother	*Moma*
Mother's elder sister	*Amnich*
Mother's younger sister	*Amnich*
Mother's elder brother's wife	*Mapo-nane*
Mother's younger brother's wife	*Mapo-nane*
Elder brother	*Tega-ate*
Younger brother	*Baya*
Elder sister	*Teg-dao*
Younger sister	*Dao*
Elder brother's wife	*Borey*
Younger brother's wife	*Borey*
Elder sister's husband	*Shakpo*
Younger sister's husband	*Shakpo*
Wife	*Naar*
Husband	*Daach*
Wife's elder brother	*Shakpo*
Wife's elder brother's wife	*Dao*
Husband's elder brother	*Tegsaya*

1	2
Husband's elder brother's wife	*Dao*
Wife's elder sister	*Boyasa*
Wife's elder sister's husband	*Aate*
Husband's elder sister	*Borey*
Husband's elder sister's husband	*Horey-feech-aate*
Wife's younger brother	*Shakpo*
Wife's younger brother's wife	*Dao*
Husband's younger brother	*Zigich Borey*
Husband's younger brother's wife	*Baiech*
Wife's younger sister	*Boyasa*
Wife's younger sister's husband	*Aate*
Husband's younger sister	*Borey*
Husband's younger sister's husband	*Borey-feechya-bayach*
Wife's father	*Roo, Aapa*
Wife's mother	*Yume-ama*
Husband's father	*Roo, Aapa*
Husband's mother	*Yume-ama*
Wife's father's mother	*Tego*
Wife's father's father	*Tete*
Husband's father's father	*Tete*
Husband's father's mother	*Tego*
Wife's father's father's father	*Kotete*
Wife's father's father's mother	*Koapi*
Husband's father's father's father	*Kotete*
Husband's father's father's mother	*Koapi*
Son	*Chhang*
Daughter	*Chime, Chimed*
Elder brother's son	*Aatyochhang*
Younger brother's son	*Bayo-chhang*
Younger brother's daughter	*Bayonchimed*
Son's wife	*Stem, Namsha*
Daughter's husband	*Chhad*
Elder brother's son's wife	*Namsha, Stem*
Younger brother's son's wife	*Namsha, Stem*
Elder brother's daughter's husband	*Chhad*
Younger brother's daughter's husband	*Chhad*
Son's son	*Spach*
Son's daughter	*Chhachach-spach*
Daughter's son	*Spach*
Daughter's daughter	*Chhachach-spach*

*APPENDIX IV**

HINDU GODS AND GODDESSES WITH THEIR LOCATION

The following is the list of most of the chief village gods of Kinnaur and it will be seen that many of the names are tribal versions of classic Hindu god names like Nag, Narain, Shankar, Shanischer, Maheshwar, Kans Raj, Gandharwa, etc. while many others defy such interpretation :

Village	Deota
Akpa	Milakyum
Asrang	Narenas
Barang	Nages
Bari	Nages
Barsering	Badri Narain (Narenas)
Baturing	Nages (of Sapni)
Brele	Devi of Kothi (Chandika or Durga)
Bruang	Nages
Chasang	Narenas
Chhitkul	Mathi
Chini	Narenas
Chugaon	Maheshras
Dabling	Dabla
Dubling	Dable
Duni	Devi of Kothi (Chandika or Durga)
Gorso	Kuldeo
Grosnam	Maheshras
Gramang	Maheshras of Bhaba
Gyabung	Tungma
Hango	Dabla
Huri	Maheshras
Jangi	Gyangmagyum
Chi-Kamba	Nages
Teh-Kamba	Ukha
Kamru	Badrinath
Kanam	Dabla
Khabo	Dabla
Shango	Maheshras of Bhaba
Khwangi	Markaring

* From T.S. Negi's, *Scheduled Tribes of Himachal Pradesh : A Profile.*

Kilba	Narenas
Koni	Nages of Sapni
Kosme	Chandika
Kostampi (Kothi)	Chandika or Durga or Devi
Krabe	Maheshras
Karla	Khoruwshoo Devi (same as Spillo)
Labrang	Chhakoling
Lambar	Rapukch of Thangi
Lippa	Tanta Narenas
Mellam	Narenas
Nachar	Ukha
Nako	Deodum
Namgya	Dabla
Natpa	Nages
Nesang	Shangche
Mebar	Nages
Miru	Nages (Swarga)
Morang	Ormig Shoo Dumber
Pangi	Sheshering
Pilo	Khormo (Khoruwshoo Devi)
Punang	Nanyo
Punnam	Pathoro (Patang)
Poari	Shankras (Parka)
Rakchhan	Shanshras
Rarang	Pathoro
Ribba	Kasurajas
Rispa	Kuldyo
Rogi	Rog-shu (Narenas)
Ropa	Chandika
Rupi	Teras
Sangla	Nages
Sapni	Nages
Shohang	Narenas
Sungra	Maheshras
Spuwa	Dabla
Sunnam	Yulsa
Shyaso	Dabla
Tailingi	Devi of Kothi (Chandika or Durga)
Taling	Dabla
Thangi	Rapukch
Tranda	Tara (Chiter Rekha)
Urni	Narenas
Yangpa	Maheshras of Bhaba
Yana	Gandrapas
Yote	Maheshras of Bhaba
Yula	Narenas
Ywaring	Chandika or Durga, Devi of Kothi

APPENDIX V

FAIRS AND FESTIVALS WITH THEIR LOCATION AND DURATION

Fair/ Festivals	Village	Tahsil/ Sub-tahsil	Dates and duration
	Chaitra (March-April)		
Chatrol	Chini	Kalpa	Either on Tuesday or Saturday during the first week of the month
-do-	Khawangi	-do-	In the bright half of the moon
-do-	Kothi	-do-	-do-
-do-	Pangi	-do-	One day (on *Amawas* or new moon)
Chishu	Sumra	Leo	One day
Shupyat	Jangi	Morang	One day during dark half
Sarimating	Morang	-do-	One day during bright half, no fixed date
Chatrol	Rarang	-do-	One day (fourteenth, bright half)
Jithu	-do-	-do-	One day, on 7th Chaitra
Rokshu Puja	Ribba	-do-	One day, on Saturday
Chatrol	Chagaon	Nachar	On full moon day
-do-	Kanam	Puh	Occasionally in the month
-do-	Kamru (Mone)	Sangla	Two days, on Saturday or Tuesday during *Navratras*
Navratra	Rakchham	-do-	One day on 1st *Navratra*
Chatrol	Sangla	-do-	For two days, no fixed date
	Vaisakha (April-May)		
Beesh	Barang	Kalpa	Two days, on 1st and 2nd

Beesh	Chini	-do-	On 1st
-do-	Duni	-do-	Two days, on 1st and 2nd
-do-	Khawangi	-do-	On 1st
-do-	Kothi	-do-	-do-
-do-	Pangi	Kalpa	-do-
Phloring	-do-	-do-	One day (after flowering of chuli and baimi)
Beesh	Pawari	-do-	Two days, on 1st and 2nd
-do-	Roghi	-do-	On first
-do-	Telangi	-do-	Two days, 1st and 2nd

Jyaistha (May-June)

Urgin Gyalja	Hango	Leo	Two days (during bright half)
Laitahsi	-do-	-do-	One day, following Urgin
Beesh	Jangi	Morang	Two days, on 1st and 2nd
-do-	Lippa	-do-	-do-
-do-	Morang	-do-	-do-
Kulang Shuyan	-do-	-do-	One day
Saringkora	Nesang	-do-	-do-
Beesh	Rarang	-do-	One day, on 1st
-do-	Ribba	-do-	Two days, on 1st and 2nd
-do-	Rispa	-do-	One day, on 1st
Remnes	-do-	-do-	On full moon day
Beesh	Thangi	-do-	One day, on 1st
Bishu	Chagaon	Nachar	-do-
Beesh	Chauhra	Nachar	One day, on 1st
-do-	Catgaon (Gramang)	-do-	Two days, on 1st and 2nd
-do-	Natpa	-do-	Three days, 1st to 3rd
Bishu	Rupi	-do-	-do-
Beesh	Giabong	Puh	One day, on 1st
-do-	Kanam	-do-	Three days, 1st to 3rd
Labrang Zalma	-do-	-do-	For one day
Giza	Puh	-do-	Seven days, during bright half
Beesh	Ropa	-do-	One day, on 1st

Beesh	Sangnam	-do-	Five days, from 1st
-do-	Kamru (Mone)	Sangla	Three days, from 1st
-do-	Kilba	-do-	One day, on 1st
-do-	Rakchham	-do-	-do-
-do-	Sangla	-do-	-do-
-do-	Shaung	-do-	Two days, on 1st and 2nd
Raskayang	—	—	Throughout Kinnaur
Purchun	Khawangi	Kalpa	In the bright half of the moon
-do-	Kothi	-do-	-do-
Jeshtang Chhecha	Pangi	-do-	One day (10th of bright half)
Ramnes	Pawari	-do-	One day on full moon day
Jeshtang Songa	Roghi	-do-	One day on 15th
Chishu	Sumrah	Leo	One day
Pita Paza	Morang	Morang	One day
Giza	Nesang	-do-	One day, on full moon day
Dumgyurja	Rarang	-do-	One day (Buddha Purnima)
Ramnes	Ribba	-do-	Two days, on 14th and 15th of bright half
-do-	Rispa	-do-	One day, on full moon day
Jeshtang Songa	Thang i	-do-	One day, no fixed date
Airatang	Chagaon	Nachar	On twenty-fifth
Jeshtang Sankranting	Katgaon (Gramang)	-do-	One day, on 1st
Sankranting	Natpa	-do-	-do-
Jangyachen	Giabong	Puh	One day
Labrangjalma	Kanam	-do-	-do-
Jangyachen	Ropa	-do-	-do-
Rola Chhango	Kamru (Mone)	Sangla	Every night from 1st to 15th
Jeshtang Songa Sangla		-do-	. . .

Asadha (June-July)

Ashletang	Kothi	Kalpa	One day, during bright half

Jeshtang (Sankranting)	Telangi	-do-	On 1st
Bumkar	Shyalkhar	Leo	One day, on full moon
Dumgyur Jalkha	Jangi	-do-	One day, on 12th
Kangyur Jalkha	Lippa	Morang	Two days, on 1st and 2nd
Ramnes	Morang	-do-	One day, on 8th or 9th of bright half
Nazal	Nesang	Morang	One day, on 10th of bright half
Ramnes	Ribba	-do-	One day, during bright half
Ponasing	Rispa	-do-	Two days, on 14th and 15th of bright half
Bhojang	Rarang	-do-	One day (date determined by the deity)
Bagh Jatrang	Thangi	-do-	One day, no fixed date
Ashletang	Chagaon	Nachar	On full moon day
Surpu Gunfa Jalma	Kanam	Puh	One day, on 7th
Kangyur	-do-	-do-	One day, on 15th
Dumgyour Jalma	Spilo	-do-	. . .
Chhokten Jalma	Labrang	-do-	. . .
Rangjun	Puh	-do-	Two days
Giza	-do-	-do-	. . .
Kailash Jalma	Spilo	-do-	. . .
Shushtan	Kamru (Mone)	Sangla	One day, on 14th of bright half
Marja	Rakchham	-do-	Two days, commencing either on Saturday or Tuesday

Sravana (July-August)

Udaneyang	Brelangi	Kalpa	. . .
Dakhrain	Chini	-do-	On 1st
-do-	Khawangi	-do-	-do-
Udaneyang	-do-	-do-	In the dark half of moon

Dakhrain	Kothi	Kalpa	On 1st
Udaneyang	Kothi	-do-	In the dark half of moon
Shonetang	-do-	-do-	One day, during bright half
Koshme Shukud	-do-	-do-	Date is fixed by Chandika Devi
Dakhrain	Pangi	-do-	For three or five days
-do-	Pawari	-do-	One day, on 1st
-do-	Roghi	-do-	-do-
-do-	Telangi	-do-	-do-
Namganchuan	Chango	Leo	One day
-do-	Shyalkhar	-do-	Two days
Dakhrain	Jangi	Morang	Two days, on 1st and 2nd
-do-	Lippa	-do-	-do-
-do-	Morang	-do-	One day, on 1st
Rangkoching	-do-	-do-	On 14th and 15th of bright half
Dakhrain	Rarang	-do-	Three days, from 1st
-do-	Ribba	-do-	One day, on 1st
-do-	Rispa	-do-	-do-
Sawanang Santhang	Thangi	-do-	Throughout the month on alternate nights
Dakhrain	Chagaon	Nachar	On 1st
Dakhlaini	Chauhra	-do-	One day
Shonetang	Kashpo (Nachar)	-do-	One day, on 15th
Dakhrain	Katgaon Gramang	-do-	One day, on 1st
Shonechang	-do-	-do-	Seven days, during bright half
Dakhrain	Natpa	Nachar	One day, on 1st
-do-	Rupi	-do-	Three days, from 1st
-do-	Kanam	Puh	One day, on 1st
-do-	Kamru (Mone)	Sangla	Two days, from Saturday or Tuesday
-do-	Kilba	-do-	One day, on 1st
Shonetang	Rakchham	-do-	One day, on 20th
Dakhrain	Sangla	-do-	. . .
-do-	Shaung	-do-	Two days, no fixed date

Bhadra (August-September)

Ukhyang	Barang	Kalpa	Five days, commencing on 19th

Penssing	Chini	Kalpa	On 7th
Jagro	-do-	-do-	On 20th
Homang	Khawangi	-do-	. . .
-do-	Kothi	-do-	. . .
Jagro	Pangi	-do-	Two days, on 20th and 21st
Panasing	-do-	-do-	One day, during bright half
Bhadrang Songa	Pawari	-do-	One day, on 15th
Koshme	Telangi	-do-	During first week
Namgan	Hango	Leo	Five days, from 18th
Manthoke or Jogtog	-do-	-do-	. . .
Namgan	Leo	-do-	Five days, commencing from 18th
Namganchunan	Nako	-do-	Four days
Namganchheya	-do-	-do-	Two days on 27th and 28th
Namganchuan	Sumra	-do-	Three days
Rangkorang	Jangi	Morang	Two days (once in three years)
Changmang	Lippa	-do-	. . .
Jagro	Morang	-do-	Two days, on 4th and 5th
Niza	Nesang	-do-	One day, on 20th
Ponas or Ponasing	Rarang	-do-	One day, on 14th of bright half
Jeu Ukhyang	Ribba	-do-	One day, on 10th
Niza	-do-	-do-	One day, on 20th
Ukhyang	-do-	-do-	From 28th
Niza (Jagro)	Rispa	-do-	One day
Rangkorang chim	Thangi	-do-	Two days, between 5th and 10th
Ponasing	Chagaon	Nachar	On full moon day
Chauhra	Chauhra	-do-	One day, on 17th or 18th
Ponasing	Katgaon	-do-	One day, on full moon day
Ukhyang	Punang	-do-	Four days, from 20th
-do-	Ramni	-do-	-do-
-do-	Rupi	-do-	Five days, from 7th
-do-	Giabong	Puh	Two days on 20th and 21st
Menthoko	Kanam	-do-	One day, on 19th or 20th

Shuktuk	Namgya	Puh	Three days, from 14th
-do-	Puh	-do-	Two days
-do-	Ropa	-do-	Two days, on 20th and 21st
Ukhyang	Rushkalang	-do-	Five days from 20th
Menthoko	Sanganam	-do-	Five days, from 20th
Janam Ashtmi	Kamru	Sangla	One day
Ukhyang	-do-	-do-	Three days, from 19th
-do-	Kilba	-do-	Four days, from 20th
Usko (Ukhyang)	Rakchham	Sangla	Four days, from 20th
Ukhyang	Sangla	-do-	. . .
-do-	Sapni	-do-	Eight days, from 16th
-do-	Shaung	-do-	Three days, from 19th
Jagro	-do-	-do-	One day, on Tuesday or Saturday

Asvina (September-October)

Ukhyang	Chini	Kalpa	On Dussehra day
-do-	Khawangi	-do-	Five days, from 10th of bright half
-do-	Kothi	-do-	-do-
-do-	Pangi	-do-	Three days, from Dussehra
-do-	Pawari	-do-	Three days from 1st
-do-	Roghi	-do-	Three days, after Dussehra
-do-	Telangi	-do-	On Dussehra day
Namgan Chhetpo	Shyalkhar	Leo	One day
Namgancha	Sumra	-do-	Two days
Ukhyang	Jangi	Morang	Four days, from 20th
-do-	Morang	-do-	Seven days, from 1st
Shirkan	Nesang	-do-	-do-
Ukhyang	Rarang	-do-	Three days, from 20th
-do-	Rispa	-do-	Five days from 1st
-do-	Thangi	-do-	-do-
-do-	Chagaon	Nachar	25th to 31st
-do-	Chauhra	-do-	Three days, from 20th
-do-	Natpa	-do-	-do-
Namkyangma Ukhyang	Kanam	Puh	. . .
Ukhyang	Kamru (Mone)	Sangla	Three days from 19th

Dang Ukhyang	Kamru	Sangla	Two days, no fixed date
Jagro	Rakchham	-do-	One day, on 20th
Gato Ukhyang or *Dang Ukhyang*	Sangla	-do-	...

Kartika (October-November)

Diwal	Roghi	Kalpa	One day, on *Amawas*
Namgancha	Chango	Leo	Three days
Rofu	Hango	-do-	...
Ganga fair	Changmang (above Lippa)	Morang	...
Shurgurich	Jangi	-do-	Two days, on 10th and 11th
Ukhyang	Lippa	-do-	Three days, from 3rd
Jai Ukhyang	Morang	-do-	Two days, on 1st and 2nd
Ukhyang	Jani	Nachar	Three days, from 1st
-do-	Katgaon	-do-	Seven days, from 10th
-do-	Glabond	Puh	...
Shirkan	Kanam	-do-	Five days, from 7th
-do-	Namgya	-do-	Three days, from 7th
-do-	Puh	-do-	Seven days, from 1st
Ukhyang	Ropa	-do-	...
Shirkan	Rushkalang	-do-	Five days, from 1st
Jagro	Kamru (Mone)	Sangla	Two days, no fixed date
Rig Diwal	-do-	-do-	One day, on *Amawas*
-do-	Sangla	-do-	...

Agarahayana (November-December)

Diwal	Barang	Kalpa	...
-do-	Chini	-do-	On *Amawas*
Shogch	-do-	-do-	This takes place in annual turns
Rathin	-do-	-do-	-do-
Pathoran Samyan	-do-	-do-	-do-
Diwal	Khawangi	-do-	Two days, during dark half
Khaning	-do-	-do-	Two days
-do-	Kothi	-do-	-do-
Diwal	Pangi	-do-	Five days, during dark half
Khaning	-do-	-do-	One day, (date determined by Chandika of Kothi)

Diwal	Pawari	Kalpa	One day, on 15th of dark half
-do-	Telangi	-do-	In the dark half of the moon
Losar	Chango	Leo	Three days
-do-	Shyalkhar	-do-	On 13th and 14th of bright moon
Diwal	Morang	Morang	Two days, from *Amawas*
Sholing	Rarang	-do-	Three days, from 13th
Diwal	-do-	-do-	One day before *Amawas*
Khepa	-do-	-do-	One day
Diwal	Ribba	-do-	11th of dark half
-do-	Rispa	-do-	One day, on 15th of dark half
Duiyal	Thangi	-do-	One day, on 15th of dark half
Diwal	Chagaon	Nachar	On *Amawas*
Deoli	Chauhra	-do-	One day, (date determined by Sungra Maheshras)
Manu	Katgaon	Nachar	One day, (date determined by Maheshras of Bhabha)
Diwal	Napta	-do-	One day, no fixed date
Deoli	Rupi	-do-	Three days, (date determined by the deity)
Diwal	Sungra	-do-	One day, (date determined by the local deity)
Khwangi	Kanam	Puh	Three days, from 5th of the bright half
Khepa	-do-	-do-	Three days, from 9th of dark half
Diwal	Kamru	Sangla	One day, (one month after Diwali)
-do-	Kilba	-do-	One day, on *Amawas*
-do-	Rakchham	-do-	-do-
-do-	Sangla	-do-	. . .
-do-	Shaung	-do-	One day, on *Amawas*
	Pausa (December-January)		
Rathin	Chini	Kalpa	. . .
Khepa	Khawangi	-do-	Two days, on *Amawas*, and a day earlier

Khepa	Kothi	Kalpa	One day, (date determined by the Chandika of Kothi)
-do-	Pangi	-do-	-do-
Parkaoning	Pawari	-do-	One day, no fixed date
Khepa	Roghi	-do-	Commences either on Tuesday or Saturday in the dark half of moon
-do-	Telangi	-do-	On the date determined by the Chandika of Kothi
Losar	Hango	Leo	Three days, from 1st
-do-	Leo	-do-	-do-
Losar	Noko	Leo	Three days
-do-	Sumra	-do-	-do-
-do-	Jangi	Morang	Three days, from 9th of bright half
-do-	Lippa	-do-	-do-
-do-	Morang	-do-	Two days, on 7th and 8th of bright half
Khepa	-do-	-do-	On 7th and 8th of the dark half
Losar	Norang	-do-	Six days, from 13th of the dark half
-do-	Rarang	-do-	Two days, on 9th and 10th
Khepa	Rispa	-do-	Three days, from 15th of bright half
Yungnas or *Jungnas*	-do-	-do-	. . .
Losar	Thangi	-do-	Two days, on 7th and 8th of bright half
Ponasing	Chagaon	Nachar	On full moon day
Ragul	-do-	-do-	-do-
Khepa	Katgaon (Gramang)	-do-	No fixed date
Losar	Giabang	Puh	Five days, during bright half
-do-	Kanam	-do-	-do-
-do-	Namgya	-do-	Three days, during bright half
-do-	Puh	-do-	Seven days
-do-	Sangnam	-do-	-do-
Khepa	Kamru	Sangla	Three days, from 14th of the dark half

Khepa	Rakchham	Sangla	One day, on *Amawas*
-do-	Sangla	-do-	One day
-do-	Shaung	-do-	One day, on 14th of the dark half
Aitang (Agitarang)	Rispa	Morang	One day, during dark half, no fixed date

Magha (January-February)

Sazo	Chini	Kalpa	On 1st
-do-	Khawangi	-do-	-do-
Mang Sanga	-do-	-do-	One day, on 15th
Jagang	-do-	-do-	One day, during bright half
Ratingahukun	-do-	-do-	One day, (date determined by Chandika of Kothi)
Sazo	Kothi	-do-	One day, on 1st
Mang Songa	-do-	-do-	One day, on 15th
Jagang	-do-	-do-	One day, during bright half
Ratingahukun	-do-	-do-	One day, (date determined by Chandika of Kothi)
Sazo	Pangi	-do-	One day, on 1st
-do-	Pawari	-do-	Two days, on 1st and 2nd
-do-	Roghi	-do-	One day, on 1st
-do-	-do-	-do-	-do-
Lamoche	Hango	Leo	Three days
Yum	Leo	-do-	Seven days, from 8th
Ngana	Nako	-do-	Four days, from 13th of bright half
Rofu	-do-	-do-	Three days, (end of *Magh*)
Guru	Sumra	-do-	Three days, from 10th bright half
Sazo	Jangi	Morang	One day, on 1st
-do-	Lippa	-do-	-do-
Mangh Shirang	-do-	-do-	Two days
Urmig	Morang	-do-	Eight days, from 10th of bright half
Lamoche	Nesang	-do-	Seven days, (after a month of Tapsya)
Sazo	Rarang	Morang	One day, on 1st
Suskar	Ribba	-do-	One day

Sazo	Rispa	Morang	As at Morang
Mang	-do-	-do-	. . .
Mangh Santhang	Shangi	-do-	Eight days, from 10th of bright half
Sazo	Chagaon	-do-	On 1st
Mangh Songa	-do-	-do-	On 15th
Sazo	Chauhra	Nachar	Two days, on 1st and 2nd
-do-	Katgaon (Gramang)	-do-	One day, on 1st
Mangh Songa	-do-	-do-	One day, on 15th
Shujarch	Rupi	-do-	Twenty days, no fixed date
Mangh Songa	Kanam	Puh	Five days, from 15th
Sazo	-dō-	-do-	One day, on 1st
Jakhya Kushimig or *Jokhya Chhugshimig*	-do-	-do-	Five days
Sazo	Kamru	Sangla	One day, on 1st
Jagang	Kilba	-do-	Three days commencing on Tuesday or Saturday during bright half
Mangh Songa	Rakchham	-do-	One day, on 15th
Sazo	Sangla	-do-	One day
-do-	Shaung	-do-	One day, on 1st
Mangh Songa	-do-	-do-	One day, on 15th

Phalguna (February-March)

Suskar	Chini	Kalpa	In the bright half of the moon
-do-	Khawangi	-do-	For fourteen days, one month after jagang
-do-	Kothi	-do-	-do-
-do-	Pangi	-do-	For eight days, (date determined by Chandika Devi)
-do-	Pawari	-do-	Eight days, no fixed date
-do-	Roghi	-do-	on 7th of bright half of moon

Suskar	Telangi	Kalpa	On a day fixed by the Chandika Devi
Rofu	Chango	Leo	One day
Laitashi	Nako	-do-	-do-
Jithu	Sumra	-do-	Three days
Lamoche	Jangi	Morang	One day
Urmig	Lippa	-do-	Eight days, from 7th of bright half
Shupyat	Morang	-do-	Four days, from 14th of dark half
Khepa	Nesang	-do-	Two days, in bright half
Jagang	Rarang	-do-	One day, on 12th or 13th of bright half
Lamoche	-do-	-do-	Two days, no fixed date
Shupyach	-do-	-do-	Three days, no fixed date
Lamoche	Ribba	-do-	One day
Phagul	Chagaon	Nachar	In the bright half starting from Monday and lasting for six days
Shivrat	-do-	-do-	On *Amawas*
Phagul	Chauhra	-do-	Six days, (date determined by Nachar Ukha)
-do-	Katgaon (Gramang)	-do-	Nine days, (date determined by Maheshras)
-do-	Natpa	-do-	Five days, no fixed date
Phaguli	Rupi	-do-	Four days, (date determined by the deity)
Suskar	Giabong	Puh	Five days, during bright half
Chunglong	-do-	-do-	Two days, on 20th and 21st
Lamat	Kanam	-do-	Three days, from 4th of bright half of the moon
Lamoche	Namgya	-do-	Four days, during bright half of the moon
Suskar	Ropa	-do-	Eight days, during bright half
Phagul	Kamru (Mone)	Sangla	Twelve days
-do-	Sangla	-do-	. . .
-do-	Shaung	-do-	Commencing on Holi for five days

*APPENDIX VI**

GENEALOGY AT KAMRU

1. Parduman Singh
2. Chhubal Singh
3. Ser
4. Kamal
5. Gulab
6. Vardev
7 Mehrup
8. Hari
9. Sarjit
10. Jagbir
11. Raghu
12. Gopal
13. Hari Charan
14. Makarman
15. Mudai
16. Bhup
17. Umed
18. Harkar Pal
19. Karpal
20. Har Dev
21. Salab
22. Bima
23. Bagal
24. Purva
25. Mehar
26. Sabla
27. Hami
28. Jawar
29. Gavardan
30. Jagbir
31. Surjan
32. Madan
33. Gobind
34. Pritam
35. Gurdari
36. Kisan
37. Visan
38. Raghunath
39. Devi
40. Charan
41. Padesri
42. Malbahadar
43. Gopi
44. Gurbadal
45. Jagat
46. Amrit
47. Dalbadar
48. Neil
49. Haripad
50. Fateh
51. Amar
52. Mahabadar
53. Salar
54. Jagbe
55. Jogdeyal
56. Dalab
57. Mador
58. Dalip
59. Jagtav
60. Guman
61. Parmodh
62. Mahipar
63. Sarv
64. Salehi
65. Gorkokal
66. Pardevar
67. Varpal
68. Charmed
69. Darjod
70. Darkori
71. Pritam
72. Sagar
73. Ran
74. Dhirjamehar

---

* From *Gazetteer of Kinnaur*.

75. Mangal
76. Gorsi
77. Lakhi
78. Prabhubhajan
79. Duman
80. Dankrit
81. Daldin
82. Pardeu
83. Mari
84. Amlar
85. Dhari
86. Basath
87. Karam
88. Prem
89. Dast
90. Charan
91. Birvesi
92. Kesari
93. Parjit
94. Dharam
95. Kamal
96. Chhatar
97. Amar
98. Karal
99. Tapnath
100. Sagram
101. Suraj
102. Darmorat
103. Charmal
104. Jabala
105. Gvasdal
106. Amrit
107. Sar
108. Krishan
109. Hari
110. Jabar
111. Bhup
112. Kalyan
113. Kehri
114. Vijay
115. Udai
116. Ram Singh
117. Rudar
118. Ugar
119. Mahinder
120. Samesar
121. Padam
122. Vir Bhadra Singh

Genealogy at Rampur

1. Parduman
2. Anurudh
3. Jamal
4. Nahar
5. Kamal
6. Jagat
7. Burid
8. Surat
9. Narje
10. Sarjit
11. Juginder
12. Raghu
13. Gopal
14. Haricharan
15. Badama
16. Budhipati
17. Bhavani
18. Ran Badal
19. Padam
20. Gurban
21. Nardev
22. Suraj
23. Bhim
24. Surmangal
25. Surma
26. Mehar
27. Jamal
28. Gajpati
29. Jawahar
30. Govardhan
31. Jagbarat
32. Surgian
33. Madan
34. Garjan
35. Jabiv
36. Girdhari
37. Kishan
38. Krishan (Visan)
39. Raghunath
40. Devi
41. Charan
42. Parmeshwar
43. Dalbadal
44. Gujrao
45. Garbadal
46. Jagat
47. Anirudh
48. Balbadur
49. Bhagwan
50. Hari
51. Amar
52. Madbhar
53. Ranmar
54. Jagpati
55. Joginderpal
56. Dalpati
57. Budhwan
58. Dalip
59. Jagpati
60. Tan
61. Narmoh
62. Manihar
63. Nardev
64. Narsingh
65. Gurubhagat
66. Mardhan
67. Nardal
68. Dev
69. Darjodhan
70. Dhenugaj
71. Pritam
72. Sar
73. Ratan
74. Dhajbhor

75. Mangal
76. Sursen
77. Bhabhi
78. Haribhajan
79. Dhanbharat
80. Bharat
81. Halsen
82. Nardev
83. Sar
84. Amar

The Genealogy of the last Twelve Rulers is as under

120. Chhatar Singh
121. Kalyan Singh
122. Kehri Singh
123. Vijay Singh
124. Udai Singh
125. Ram Singh
126. Rudar Singh
127. Ugar Singh
128. Mahindar Singh
129. Shamsher Singh
130. Padam Singh
131. Vir Bhadra Singh

APPENDIX VII

TRANSLATION OF A PAPER RESPECTING THE TRADE BETWEEN BUSSAHIR AND CHINA COMMUNICATED BY MOONSHIE CURRUMOODDEN*

Garroo a town of China is situated 180 coss East of Rampour, people repair thither from Rampour via Cunnoheer for purposes of trade, and are there met by and traffic with the Chinese of all parts. The produce of China is brought on large sheep capable of carrying from 20 to 30 seers cutcha. That of Bussahir on men's backs.

May and June are the months in which people repair to Garroo from Bussahir and June and July those in which they return. Excepting by Garroo there is no other direct communication between Bussahir and China that by Ladack being greatly circuitious.

In some instances the people of Bussahir go on from Garroo to Jhukoun a manifactory of salt and to wooshing beyond the latter they never proceed.

The Chinese resort from Garroo to Rampoor Bussahir in the month of October and return in November and December, the road is perfectly safe. But one cannot trade singly on account of the distance that something intervenes between villages or human habitations and the consequent impossibility of aid in the event of sickness or accidents.

The usual mode is to form parties for the journey of 20 or 25 men, which find accommodation in caves and excavations in the rocks sometimes capable of containing 50 men.

Garroo—though named a town does not contain a single permanent house or hut, the neighbourhood produces neither wood nor straw, and all those who resort theither dwell in tents.

*Foreign Deptt. SECRET. 27th October 1815, No. 39-42, pp. 29-35.

The country moreover produces no grain nor provisions of any description on which account the traders carry with them all that may be necessary during their stay.

The road from Jhukown to Rampour Bussahir is 1½ months march as far as such it is exceedingly bad but thence to Garroo it is practicable for horses—a red blanket is the usual dress of the people of Garroo.

ROUTE FROM RAMPOUR TO GARROO, RAMPOUR TO HATTEE . . . 6 COSS

Malta	7
Chapootah	6
Paivady	6
Aungtoo	6 coss the futby
Lassur	6
Raggy	8
Pangy	9
Leepah	10
Sundum	10
Hangoo	8
Changoo	10
Soomuth	6 frontier of China
Mancha	6 the people here dwell in tents
Dangoo	5 coss uninhabited
Budkoo	8 4 coss asent and 4 descent on the hills so-called
Poonoo	8 uninhabited
Chamoorty	6 here is a post of Chinese Troops consisting of a sub and sepahees.

To a Ferry over a River

Named

Kargool	6
Doorpind	7 uninhabited
Surtee	5 hence a road which Tattoos can travel

Rabua	4
Kanchun	5
Jurghang	6
Laboojee	4 uninhabited
Chahalug	5 a post of sub, and 80 sepahees
Jumghang	4 uninhabited
Duug	4 Ditto
Khaggeahnon	3
Hingum river	5
Garroo	5 here are stationed 2 soobas and 1000 sepahees.
Total coss	194

Trade of Bussahir

Salt is imported from Jhakown via Garroo to Bussahir. The people of Kunnohur take it in barter for Drug (a sort of raisin) and wheat at equal quantities.

Salt when brought from Kunnohur to Rampour sells per seer for 4 seer rice, 8 seers Mandooa Attah—and in money at 16 or 20 seers per rupee.

Sohagah is imported from Johakun to Bussahir and is bartered for 4 seers wheat or 4 seers drug per 5 seers of Sohauga, at Rampour it sells 7 or 8 seers per rupee.

Goats wool is sold to traders from Bussahir at Garroo at 12 seers per rupee—at Rampour it sells for 7 per rupee. Chaor Cowtails sell in Bussahir at 2,3,4,5,6,7,8,9,10, rupees each according to their quality.

Nerbussee (a Drug) is brought from Napaul to Garroo where it sells at 4 pice weight per rupee. It is imported thence into Bussahir when it sells at 3 pice weight per rupee.

The goats producing the Shawl Wool are sold to the traders from Bussahir at 2 rupees each at Garroo by whom they are resold at Kunnohur for 3 rupees each. A box of teen containing 10 seers sells at Garroo at 10 rupees.

Laddack

Laddack is distant from Bussahir 260 coss (27 days march). Traders resort from Laddack to Bussahir in the month of

February and March and they return in June and July. The road from Bussahir to Laddack is for the first 12 days march very bad but afterwards improves. The chief of Laddack pays tribute in equal proportions to China and Cashmere.

ROUTES

From Rampour to Huttee 6 Coss

Mathala	6	Auckehee	7
Toorajee	6	Soomur	7
Auktoo	10	Roogoo	6
Rojolla	6	Chohoo	8
Roonghee	4	Gunpau	8
Phaungee	7	Tokechu & Chumn	10
Lepee	10	Juglug	7
Hill Somum	10	Gaoo	6
Hanggoo	7	Mer	5
A Nolla	6	Gundy	7
Chunggo	6	Himmee	8
Sucktoo	10	Sugnah	6
Purgaoo	7	Chock Seth	7
Ghary	5		

CITY SEOO

The country near Seoo is very productive and silver is found there. It is purchased there for the seer and sold at Bussahir for the Mahomed Sahee rupees. The comparative value of these two being as 4 to 5 a fifth is thus added to its price. Numdah Borteedar 5 hauts long and 3 hauts broad is purchased at Seoo for 4 and 5 sells at Rampour at 7 and 8 rupees.

Saffron is purchased at Seoo at 10 rupees per cutcha seer.

Articles produced in Bussahir 2 rupees per tola Musk.

Hauk (Bauz)————100 rupees

Ditto (joora)————60 Ditto.

Bhojeputtur

Daukh is purchased at Cunnohur at 25 and 30 and sold at Rampour at 16 and 20 seers per rupee.

Opium 5 seers p. 10 Rs.
Churuss 5 (do p.8)
Mohra (a poisonous drug)

Iron pucka 4 Rs. 8 p. maund cucha 2 Rs. p. do; white pungkee (woollen stuff) from 3 to 6 Rs. p. piece. Sucklad (Woolen stuff) 5 to 6 Rs. each, Rock salt brought from Mundee sells at 1 maund 10-seers per rupee.

There are two annual fairs one in October the other in June at Rampour whether people resort from all quarters, and where every thing is to be found for sale.

Articles imported to Rampour from Hindoostan

Tobacco-sells at Rampour at 12 seers per rupee and is bartered for 3 seers wool per seer Tobacco.

Googul/perfume/
White taul/wax/
Jhola cloth
Goor
Salt Lahoree
Coarse chintz
Musshroo
Bofta Cloth
Dates
Almound
Khopra Coconut
Coconut shells for
Smoking
Cardamoms

Cloves
Nutmeg
White Sundul
Koosoom/red dye/
Turmeric
Ginger
Assafatuald
Soap
Red wool
Coarse thread
Roop Lead
Tin and Pewter Sugar/candy/
coarse/
Coarse pegar
Indigo
Swords and shields.

WOOSHUNG

A country named Wooshing by N. E., from Rampour distant 3 months March from Rampour via Garroo Hither a few people have occasionally repaired from Bussahir generally in disguise of Fakeers, but never further—there is no danger however on the road.

True translation signed
R. Ross. Asst., A.G.G.

APPENDIX VIII

BUSSAHIR*

39. This is the most remote part of the British dominions in Asia. Its situation renders it interesting in its political, geographical and physical relations, being bounded by the dominions of China on the north and east; Ludauk north-west; and by Cooloo (trans-Sutlej) west; south by the river Nauglee, which runs into the Sutlej ten miles south of the capital, Rampour. Half of the territory lies hither Himalaya and the remainder, which is known by the name Kunawur, within this snowy barrier.

Description of Bashahr

Boundaries

The grand divisions of this Principality may be taken as under:

Grand divisions

Dusso	Includes that portion contained in the valley of the Sutlej up to Sooran.
Kunawur	Commencing immediately above Pooaree and lies within the snowy crest of the Himalaya.
Chooara	Occupies both banks of the River Pauber and its tributary streams.

Foundation, Sumbat 472: This Raj was founded by an emigrant Rajpoot from the Deccan in Sumbat 472, by name Dumber Sing.

40. Annexations: Thè Thakoorais of Dulaitoo, Kurungoloo and Kunaitoo were annexed about Sumbat 1611.

*Extract from Captain Kennedy's Report on the Protected Hill State, 1824

Comharsain, Sari, Shangree and Kotgooroo were wrested from Keonthul Sumbat 1761 and annexed as tributaries to this principality.

Raen was also wrested about the same time from Gurwal.

In Sumbat 1868 (A.D. 1810) died Raja Oogar Sain. He left an infant heir, the present Raja, 18 months old, and a few months subsequent to his decease the Nepal troops took possession of this State, with the exception of Kunawur, to which place the infant Raja with his mother and the hereditary Wuzeers fled. The Nepal troops followed them as far as Wangtoo, where there was a wooden bridge, called a Sanga, across the Sutlej. A Partial action occurred between the Kunawarees and the Goorkhas, which was followed by the retreat of the latter, but not until the bridge had been destroyed by the inhabitants of the country to cut off the communication with Bussahir proper, which remained in the hands of the Goorkhas.

The Bussahir Prince with his mother remained in, Kunawur, but they paid Rs. 12,000 tribute per annum, for being left in peace, to the Goorkhas, until the advance of the British troops, when they returned to the capital in Bussahir.

Revenue of Bashahr

41. The revenue of this province at the period of invasion of the Nepalese, as taken by Captain Ross, amounts to Rs. 67,000 per annum.

42. At the period of British conquest of these mountains, little or nothing was known of this portion of the territory, particularly that lying within the snowy range; and although the value of the soil was ascertained, yet its extent and boundaries seemed to indicate far greater

resources, which actual inspection and survey only proved to be imaginary.

Travellers and men of science behold the singular spectacle of a country granded by natural and almost impregnable barriers, which also formed the country itself and upon sites where labour was scarcely productive, the loftier regions and summits repelling every nutritious particle and veiled in central snow (sic).

Prosperous condition of the people

43. Scanty as the population is of the interior of this province and unproductive the soil, the inhabitants are perhaps more comfortable and in better circumstances; they certainly possess a greater freedom than in any other district in these hills.

Surrounded by rocks of difficult access, inimical to industry, offering little inducement to cultivation, in a climate of protracted rigors, which nature dictates, and we find them active, enterprising and industrious, occupied in extensive commercial intercourse and trade; rearing vast flocks which form their chief dependence and trafficking into remote countries under great hardships and privation to gain a comfortable subsistence for their families at home.

Absence of Crime

44. In no part of the protected dominion, and I may give a wider scope and say the world, is there less crime known.

Petty Chiefs

45. Independent of their hereditary Wuzeers in this province, every pargana has a chief with that title annexed to his name.

Aggressions into Bashahr and retailation

46. Bussahir, notwithstanding its physical strength, has by its annexations with State of greater resources occasionally been visited by aggression and has become the aggressor in return. From west to south it has no powerful neighbour, being there in contact with the small States of Comharsain, Kotgooroo, Joobul, Poonidur, and Gurwal, none of which,

with the exception of the latter, possesses mean of aggrandizement and the government of this State is too weak and vacilating to undertake systematic warfare.

The frontier borders, towards its northern and eastern frontier, who own little allegiance to their Chiefs, frequently made irruptions in former periods into Bussahir, plundered and laid waste the territory, and retreated again to their wilds, resuming their aggressions on the first impulse of their predatory habits.

The valley of the Pauber has been the chief scene of their depredations, from whence the flocks have been carried off. The frontier pargana of Neelung in the Chinese territory has been frequently plundered by these banditti.

South-east to north-east, Bussahir confines upon Tartary, all within the snowy mountains; north and north-west it comes in contact with Spiti of Ludauk; and west it borders upon Cooloo trans-Sutlej, the river forming the line of demarcation. All these boundaries have in former times been the subject of dispute and contention. Tartars, from Neelung (on the Gunges) and from the great tableland beyond Shipkee have visited the country with hostile views.

The pargana of Hung Rung, the most remote portion of Bussahir, formerly belonged to the Chinese; its inhabitants are Tartars, and have the same language and customs as their neighbours who are subjects of the Chinese Government. The Ludauk frontier has been the scene of much desultory warfare, but the aggressions on either resemble those that formerly occurred in Scotland in feudal times, consisting of forays and assaults villages, etc. But these skirmishes seem never to have distracted the internal quiteness of prosperity of the country. Many years ago Bussahir got possession of

Dunker, a fort of Spiti, but its present frontier at Shealkur is more eligible for securing the tranquility of the country, and under this idea Dunker may have been relinquished. The Great barrier of snowy mountains, towards Ludauk, formidable as it is, has frequently been crossed by parties, of robbers, who after plundering the frontier villages and carrying off their flocks, passes that none but desperadoes would attempt.

Political importance of Bashahr

47. Bussahir, considered in a political light possesses considerable interest. It confines upon the Chinese territory and Ludauk, as has already been stated, and communicates through the latter territory with Yarkund and Kashgar, connexions that give it a political aspect of some importance. It may be born in mind that, while the house of Timour ruled the destinies of Hindustan, a Tartar Government managed the Empire of China. These relations give the country a geographical interest, more lively when we recollect that it sends forth the great rivers which formed the scenes and boundary of Alexander's conquests in Asia.

Interesting features of the country

48. In a physical view the extraordinary character of the neighbouring country (the great plateau of Tartary), the singular elevation of the soil, the climate, productions, and inhabitants—all points to conclusions of which we have only the signs.

49. Regions producing the shawl-wool goat, and the yak (an invaluable animal), tracts said to be rich in metallic wealth and inhabited by singular tribes, whose character and resources, we know but little of, altogether attach an intense interest to this great Central Plateau.

Description of the portion of Bashahr called Choara

50. The portion of Bussahir called Chooara (so named from its producing a reddish species of rice) lies all hither to the Himalaya range, but ramifying amongst the roots and gorges of

snowy mountains (as has been stated) on both banks of the river Pauber, which winds up amongst the peaks and spures of the snowy chain, having its source in a lake called "Churamuna" of perpetual ice upon southern slope of the Himalaya at an elevation of 13,700 feet. The ridge is crossed above the lake by three passes called Goonas, Neebrung, and Ghoosool 16,000 fit above the level of the area. The descent on the other side leads into Kunawur by the valley of Buspa. There are five nalas of divisions in Chooara which are again sub-divided into others, and were formerly under petty Chieftains.

51. The lower parts of this valley (of the Pauber) at a town called Rooraokotie (where the bed of the river is 5,000 feet above the level of the sea) are very fertile. Rice is the chief crop, the fields of which are irrigated from the river Pauber and torrents from the snow beds. The cultivation far exceeds the consumption, and the surplus is exported to Kunawur in exchange for wool, and to Nawur for salt, where it is again bartered for iron, etc., etc.

Description of Tukral and its inhabitants & c.

52. Tukral is the remotest inhabited portion of the valley approaching the source of the river (Pauber). The country is extremely wild and rugged, exhibiting steep cliffs crowned with eternal snow, and their basis clothed in deep for us. The inhabitants of this region assume the same character as their mountains, rude, savage, warlike and independent living in seclusion and naturally of a ferocious character. Only a few years since they owned but little allegiance to their Government, and the revenues could only be collected by an armed force. They were a cast of independence and self confidence which is unknown amongst Asiatics. Their country is inhospitable, climate harsh, and the soil yields them but a

scanty subsistence, and this reluctantly. Naturally savage and their hopes often blighted by the climate, they appear to live at variance with themselves. They are hardy and courageous, but given to plunder. In some respects they rise superior to the whole race of Asiatics, being ingenious and enterprising; in others, they sink into the grossest abandonment of all principles. Their weapons are chiefly bows and arrows, the latter barbed with bone. Every man's house is his castle. They seem always ready prepared to give or to resent an injury, considering, I presume, "the surest way to keep at peace is to be ever prepared to go to war". They are fond of hunting, and pursue the chase (deer and wild goats, etc.) with keenness, through the snow. They are expert at striking a mark, and are famed for the practice of some athletic or warlike exercise. The elevation of these villages is commonly about 9,000 feet above the level of the sea, and here the climate is very rude, snow falling towards the end of October and remaining in the field until the middle of April. The inhabitants get rid of the snow in spring by throwing soil upon it, as is practised in Switzerland, which absorbs the sun's rays. In harvest time, in the highly elevated districts, the people carefully avoid discharging firearms, lest a fall of snow should succeed the concussion and destroy the crops.

53. In Kunawur the people chiefly subsist by trade, and live upon their flocks. The grape, apple, and turnip are extensively and most successfully cultivated, and form a chief source of food during the winter months. Blankets form the chief export to the Chinese territory, raising, newzas, 'nuts' tobacco, rice, horse shoes, saddles, stirrups, etc., etc., are sent to Tartary.

Female Slavery

54. In no part of the hills has female slavery had so easy a purchases as in Chooara, and this partly from the scanty subsistence the country affords, the absence of morals in the people, and the inducement of metallic wealth. Now better than the inhabitants of these upland sequestered tracts know the value of a bar of silver.

Tranquility of Bashahr

55. Two famed robbers by name Ram Chander and Micha (Rob Roys) frequently made irruptions into this part of Chooara, by difficult passes in the snowy mountains, and levied black maill ; but since the British protection has been afforded to this province, the country no longer is a prey to savage banditti or agitated by internal feuds. Everybody lives in tranquility, and Bussahir receives its revenues without the intervention of an armed force, although the Government at present may be deemed pressing and rapacious in consequence of the youth of the Rajah and he being in the hand of the hereditary Wuzeers a mere puppet.

Dress of the people of Chooara

56. The people of Chooara drees in woollens, their own manufacture; their cap is worn conical and not unbecoming.

Difficulty of passes in Kanawar

57. The valley of the Pauber communicates with Kunawur by lofty passes through the snowy chain, which are numerous, and of difficult access and are encountered at cross. Their general elevation is above 15,000 feet in the region of perpetual frost and snow, but there is no check to the adventurous spirit of man. The inhabitants of Tukral now trade extensively in iron, which they take from Nawur in Bussahir.

Dress etc., of people Bashahr

58. The population of Bussahir may be estimated at 40,000 souls, but there has not been an actual census taken as yet.

The dress of the inhabitants consists of woollens at all seasons of the year, a brown hair cap, made into the form of a turban, woollen shoes, coloured stockings in the Chinese fashion, a pouch with steel and tinder; and a dog resembling the species of a New foundland in Europe generally attends a Kunawurie.

The women dress their hair plaited down to their heels nearly. The wear extremely massive pewter rings round their ankles, which gives an idea that they were not placed there for ornament; they wear a woollen petticoat and cap like the men, with the exception of the colour of the top, which is red. There features are pleasing and they appear well skilled in husbandry and in spinning wool.

Lamas in Kanawar

59. In Kunawur the Lamas or priests take an active part in the government. Their temples are magnificent buildings of stone, in the Chinese style of architecture, and generally well filled with manuscript books and brazen images. Offerings are daily made to these gods of Puwaseen, and tracts of lands are set apart for the maintenance Lamas. There are numerous nunneries in Kunawur. The nuns do not conceal themselves, but come out and solicit alms. They are in general by no means the most comely of their sex.

Minerals

60. Mines : There are several iron mines in Bussahir which are worked by the inhabitants of the country, who appear in good circumstances. Some of the galleries extend half a mile horizontally into the side of the mountain, but they are not above three or four feet wide. They use no perpendicular shafts, so that the miners are obliged to work by torch-light.

Nawur may be considered the principal mining State, and there the iron sells for 12 seers per rupee. The Bussahir Government levies a duty of $\frac{3}{4}$ths of an anna upon a load (as much as can be carried by a man), which is from 40 to 50 seers, or nearly one cwt. The place of rendezvous for the sale or barter of the iron is Sissoo, a town in the Protected Sikh plains about 34 miles from Subathoo, and there the mountaineer either receives cash or produce of the plains, such as goor, grain, tobacco, cloths, sugar, etc., etc., for his iron.

None of the precious metals are found in Kunawur or any part of Bussahir. If a vein of metal was discovered the constitution of the mountains, composed of gneiss, granite and other most hard rocks, and the impracticable nature of the country, would entirely preclude the possibility of working it to any advantage. Copper, which always contains more or less gold, is found in one part of Kunawur. The mine was worked for a short time by a miner from Cooloo, who abandoned it as soon as the ore on the surface was exhausted. A gold mine was discovered a few years ago in the Tartar dominions adjoining Bussahir, but for what reason it was immediately shut up by orders from the "Maha Cheen" is not distinctly known.

Government of Bashahr

61. The Government of Bussahir consists of the Raja and three hereditary Wuzeers. I beg leave to notice the evil that arises to these districts where these Wuzeers are hereditary.

As soon as they obtain influence over the Chief by indulging his propensities, intoxicating his senses with drugs inducing a degree of mental imbecility, then commences the abuse of men dressed in a little brief authority; money is their only god, and they do not hesitate to give ample scope to their rapacious and tyrannical propensities.

The present Raja is 14 years of age. He was placed upon the Musnud in 1821. Only a few weeks after his inauguration his mother, who had the guardianship of her son in his minority, was burnt to death, by accident or design is not quite ascertained, but it is generally supposed the latter, and at the instigation of the Wuzeers, in order that they might be better able to continue the system they had pursued of keeping the reins of government in their own hands.

The education of the Raja has been grossly neglected, and already he has the appearance of being an opium-eater. The Wuzeers allow but a very trifling sum of the revenues which they have charge of to be appropriated to the expense of the Raja's state or comfort.

Description of the Capital, Rampur

62. Rampoor, the capital, is situated on the left or eastern bank of the river Sutlej, which is about 210 feet broad here and extremely rapid; latitude 31°.27′ 'north; longitude east 77°.38′. It is 91 miles by the main road from Subathoo, and within 10 miles from the great Himalya peaks which are covered with perpetual snow. Its elevation above the level of the sea is 3,300 feet. This capital is situated in the dell of the river, at the bottom of most precipitious; lofty, and rugged cliffs from 4 to 5,000 feet, rising almost perpendicular to the town. The reverberation of the solar rays produces intense heat, and in summer stagnates in the dell night and day (sic) the thermometer rising to 100° and not unfrequently to 110° of Fahrenheit in the shade. In the winter the sun, being so long hid by the cliffs, does not heat the valley; and there rises a raw, damp and most unwholesome

atmosphere, the effect of which is observable in the persons of the inhabitants, who are mostly sallow and sickly in appearance. In the month of November the sun was not seen at Rampoor until past 11 O'clock, and it disappeared at 3 O'clock behind the hills.

The inhabitants appear very industrious and have a manufacture of coarse shawls and other woollens. There is an excellent bazar, and at three period of the year fairs are held which are attended by people from the Sikh plains, Cooloo, Kunawur, Tartary, Ludauk, and Cashmere. One occurs in the month of May, one in October, and one, called the Dhal Mela, in December. At this fair one person from every zemeendar's house in Kunawur must be present, and the whole armed, whence the name Dhal. Hatchets and battle-axes are the most common arms. There are a few matchlocks, shields, and swords. This militia march through the town at this period and are mustered before the Raja's residence, when they fire a volley.

The Raja's house is built in the Chinese style of architecture, of stone and wood, without any kind of cement. It consists of three stories, and the workmanship appears excellent and executed with great taste.

The streets in the bazar are broad and well laid out; the houses stand in squares, having an area in the centre, appearing neat and possessing great conveniences and comforts.

At the above-mentioned fairs are to be seen some English staples, and throughout the country a preference is given to English *chintz*, which are only worn by the better order of people. A gentleman from an extensive manufacturing firm in Perth visited this part of the country in 1822 and recognized a piece of *chintz*, the pattern of which he had designed.

Rampoor may be considered the emporium of these States. It is a channel of commercial communication between Chinese, Tartary, Ladauk and Cooloo. It is the resort of people of many countries, characters, and customs. The Tartar was observed putting his wool in one scale and receiving its weight in Tobacco, or coarse sugar, from the trader of the plains of India, neither of them being able to comprehend each other's language. The common steel yard was observed in general use in the fair weighing articles of trade (sic.).

The marks of Nepal oppression meet the eye constantly in the depopulated and decayed houses in Rampoor. The town being now to wear the appearance of progressive improvement, and in the course of a few years, if the present Raja retains his senses and proves a blessing to his country, this capital may be fairly expected to resume its former flourshing trade. There is an air and appearance here altogether different from the character of Asiatic towns.

Communication with the Kulu

63. The Sutlej is crossed by a Joola of rope made of grass, of very rude construction, which communicates with Cooloo.

The State of Cooloo from Rampoor appears very sterile and warlike; every peak within view is fortified. During the winter season the Sutlej is crossed on inflated buffalo hides.

64. On account of intense heat of the summer months the Raja and his court remove, to a town distant from Rampoor 22 miles, called "Soran". It lies on the slope of mountains which, immediately above, it, were covered in the month of June with snow. These are the declining summits of the parent chain Himalaya. The Sutlej rolls in a dark worn bed Soran, rendering the scenery grand and imposing. Soran is 7,200 feet above the

level of the sea, and possesses a delightful climate. The mean temperature is lower than that of London. Here again is to be seen the hand of former oppression in the demolition by fire of the houses. The crops at the level consist of barley, oqul, phaphra, chena, and Kuddoo. Number of people live under the actual projection and shelter of rocks.

The principal temple of Soran is dedicated to the Goddess "Bheema Kallee", who is styled the "Governess of Bussahir". Human sacrifices are said to have been made at the shrine of this Deota, but have been discontinued since the British conquest. The temple is attended by Brahmins, but beyond this place none of that caste are to be seen towards Kunawur, which may account for the decidedly apparent superior comfort and morality of the people.

Description of the country in the neighbourhood of Rampoor and Soran

65. The country in the neighbourhood of Rampoor and Soran and indeed it may be applied to the whole province does not exhibit a single level spot equal to the dimensions of a field of two acres, square measurement. The pathways are wholly impassale to laden horses and mules, and, with the exception of those parts where the main road has been made, there are great difficulty and risk for sheep and goats, upon which most the trade is transported, each animal carrying from 30 to 40 lbs (English weight). It is surprising to observe the regularity and order kept up by means of dogs (a species peculiar to the mountains, huge, savage animals) in these flocks as they proceed to their destination. The whole tract of country is singularly rough and difficult of access, and may be said to be made up of clusters and ridges of sharp peaks, many of them covered with perpetual snow. The inhabited regions are confined to the dells and gorges which interest

them and drain off the streams, the cultivated parts forming so small a proportion as to appear as patches or steps of stairs up the slopes of the mountains. In the interior, as we advance from Soran towards the Tartar frontier, the inhabitants assume a Chinese aspect in their persons, dress, manners, customs, and religion, subsisting chiefly on their flocks and deriving their commodities, luxuries and metallic wealth from their trade and commerce with neighbouring countries.

66. In Kunawar the inhabitants are called after their States or houses as in Scotland, and in their intercourse with other States are better known by them.

Tea Plant

67. Two species of tea plant have been found in Kunawar, in the Soongnam dell, by Doctor Gerard.

Intercourse between Bashahr and the Chinese Frontier

68. Intercourse between Bussahir and the Chinese Frontier : The intercourse is pretty extensive. The months of May and June is the season when the people of Bussahir repair to Garoo, which is the chief mart of shawl-wool on the Tartar frontier, and the Chinese resort to Rampoor in October and return in November. No danger but that of the precipitious nature of the road is known. Formerly the remote portion of Bussahir in Kunawar was possessed by the Chinese, and was given up to them. The Tartar pargana of Hung-Rung is a portion of Bussahir, and for its size is a main source of its resources. Blankets, raisins, nuts (newzas), tobacco, rice, horse shoes, saddles, agricultural instruments, fire-arms, and other produce of the hither Himalaya from the Chief exports to the Chinese territory; wool and salt the imports. The prosperity of Bussahir mainly depends upon this intercourse. The advantages appear

mutual, but the Chinese have frequently threatened to put a stop to present, although formerly Bussahir received a trifiling respect from Neelung, which is the frontier Tartar town. Bussahir is fully aware of the singular situation and importance of this commerce, upon which it is so dependent. The distance from Rampoor to Garoo, by the banks of the Sutlej, is 100 coss; to Shipkee, the frontier village, it is 120 miles. Bussahir communicates by several routes diverging from the river Sutlej to the tableland of Tartary.

Firstly : By the river Buspa, which takes the traveller over a pass 16,000 feet above the level of the sea, and brings him to Nellung on the Chinese frontier. Many people perish in attempting this route.

Secondly : A route from the Sutlej by the course of the river Teedoong, a considerable stream, 25 miles higher up the valley. The Tartar villages of Neelung and Charung occur in the dell. This road is also difficult of access. It leads to the great tableland at the back of the snowy range and communicates with Badrinath and Kedarnath, places of superstitious Hindu reverence; also with Daba, and by the Neetee pass with Kunawar, the road lying behind the Himalaya crest.

Thirdly : By the Taytoo stream, ten miles up the dell, setting out from Neesung, a Tartar village of Bussahir. This leads to Bekhur, a Chinese District.

The pass into Tartary, from hence is over a very lofty ridge. In the month of July the snow was seen drifting on it. In the vicinity of Bekhur the country is represented, by a gentleman who has visited it, as open, and the people generally ride on horseback.

The Chinese admit no European to pass this town (Bekhur)—they have never been known to offer violence to travellers proceeding there, but mildly insist on their return. The inhabitants are represented as tall, athletic, good humoured and self-confident, remarkably well dressed and accoutred with handsome head-dress of basket or wire covered with silk, and a fringe round it, crowned by a trident.

From the town of Shipkee roads diverge in all directions, and couriers travel with despatches into the interior of China. The routes lead by Garoo, on the banks and near the forks of the Indus, and by the course of the Berhampootra to Lhassa, which is the principal seat of Government on the frontier. From thence communication are made into Yarkund, Turkistan, Samarkund, Bokhara, and the Russian Frontier.

The character of the people and the suspicious vigilance of the Chinese cannot be better illustrated than by the mode of intercourse between the frontier posts and the interior, which has been derived from Puttee Ram, a well-known, respectable merchant of Bussahir, who has frequently visited those states. Regular posts traverse the country, carrying the news of events from the remoter confines into the interior with a rapidity and precision quite astonishing, and also on a scale of rigour suited to the jealous policy of the government. The despatches are tied upon the back of a courier and sealed to it. He is then mounted upon a horse. Unless accident occurs, nothing justifies the seal being broken; in consequence the rider cannot dismount until he arrives at the next stage and all his interest lies in accomplishing the distance with the utmost speed.

The same distrust of foreigners meets the traveller on the confines of India equally as in the centre of Pekin. A gentleman who visited this frontier in 1821 represents that the intelligence of his arrival spread immediately, and that every pass was guarded by parties of Tartars to prevent his further progress. Number of Tartars were seen on the great tableland in tents made of the hair of the yak; they were frank and civil, but would not hear of his proceeding another step; their lives would be forfeited if they failed to obey the orders from Lhassa.

The people of Cooloo have occasionally invaded the Chinese territory. No resistance was made, and in reply to a question the latter said "it was not worth the consideration of so large a State as the "Maha Cheen" to send a force to swallow up the smaller".

Intercourse with the Russians

69. The Russians are known on the confines of Bussahir by the name of "Oroos". Merchants resort to Leh, the capital of Laudauk, and an active trade is carried on between the Russians and Ludaukees via Yarkund. A few articles of Russian manufacture find their way to our frontier. The chief are felts, beads, coral, amber, cloths, and leather.

Tea from China and bars of silver, stamped by authority, from Yarkund, toys from Russia, may occasionally be had at the Rampoor Fair. Mandarin chopsticks, cups, and saucers are seen in Kunawar. Dresses of men of rank, some of them very costly, silks, satins of very beautiful patterns and texture come also to the same fair.

Gold coins of Belgium and Russia are occasionally seen in the Subathoo bazar, and the purity of the gold is highly esteemed.

Trade with Ladakh and Tibet

70. The distance from Rampoor to Leh is given at 25 days' journey. Traders resort from thence to Bussahir in the months of February'

March, April, and May and return in June and July. The road for the first 12 days is most precipitous and dangerous even for man, and wholly impracticable to loaded horses or mules. From Dunker northward the road is represented as pretty good, level, and quadrupeds travel without the least danger.

Ludauk pays tribute to China and Cashmere in equal proportions.

The Kunawarees who trade to the Tibet frontier pay no duties there or in Ludauk by ancient agreement. Their commerce with these countries produces more than enough to pay their contribution to Bussahir. The British Furruckabad rupees are now the principal currency in all the frontier districts.

Rivers of Bashahr

71. Rivers: The principal rivers in Bussahir are:

1. The Sutlej
2. The Pauber
3. The Buspa
4. The Teedoong
5. The Spittee or Lee

The Sutlej

72. As the river Sutlej, independent of its celebrity in ancient geography and the obscurity and fame of its source, forms the boundary of demarcation of the British territory on the west of India, I have considered it both useful and interesting to ascertain from the best sources of information the line of its course within the mountains, and sections of its level, to the extent explored by European and this has been satisfactorily done between the limits of Loodiana in the Punjab, or Sikh plains, and Shipkee, the frontier village of the Chinese Government, which is the highest point reached by following up the stream. The river has been

visited at Bekhur on the tableland of Tartary, three days' journey beyond Shipkee, which is the most eastern point that has yet been actually attained, but it has been seen from the crest of the Neetee pass, ten days' journey higher up, and its level geometrically deduced.

Lake Munsurowur (or Mapang), the supposed source, is ten days' journey above this position, and it has been visited by Messrs Moorcroft and Hearsay. This celebrated Lake is an object of high adoration by the Hindus and Lamas. I have every reason to believe that the heights as here given by ebullition and barometrical observations are nearly as correct as could be obtained by Trigonometry, and the coincidence, when both methods were used, is proof of general accuracy. Dolland's mountain barometers were used, and thermometers were of all kinds, and the observation taken free from local influence. The few barometers that were preseved from accident were verified on returning, and found not to have varied in any degree. The boiling point of water was constantly observed as a check upon the barometrical indications.

73. I shall proceed to show the sections of level of the Sutlej commencing at Loodiana.

The Sutlej (or Hesudrus of the ancients) is the only river, within British territory, that carries its course entirely through the snowy chain of Himalaya, rising in the celebrated Lake Munsurowur, agreeable to the oral accounts of numerous native travellers, particularly the Lamas. It traverses a portion of the elevated Plateau of Tartary, entering the mountains at Bekhur, latitude north 31°35′, longitude east 79°. At Shipkee it is already confined within vast cliffs of granite; it continues inbound by hoary tops for 100 miles, and little below Wangtoo (where the river has a channel of solid granite) it emerges from the snowy zone and flowing on in a deeply indented and narrow ravine for 150 miles it debouches into the Sikh or Punjab plain at Mukuwal. At Loodiana the bed of the river appears to have a positive elevation of about 900 feet according to the observations made by the late Captain Blair of the Engineers; the expense here is very considerable and the stream is consequently shallow and intersected by sand banks, and in the cold season shows an insignificant body of water.

74. In May and June the river swells from the wing of the snow in the mountains, and in August, receiving accessions from the periodical rains, reaches its maximum limit. At the town of Belaspoor, in the Rajship of Coollor, distant from Loodiana 100 miles and two days' journey within the mountains, the river was found by Mr. Moorcroft to be about 1,500 feet above sea level, which gives a medium acclivity of about 10 feet per mile in direct distance; but if the inflexions of the stream are considered, the actual rise of the soil will scarce exceed half of this, or 5 feet per mile. At Belaspoor the river is about 100 yards broad including its sandy bed, and is crossed by inflated skins. The course of the stream hence is very devious, making sharp turns round the bases of the mountains, and is often very much contracted and jagged, without any shore of slip of soil at the margin, but washing the feet of the mountains.

75. Near Soonee, in Bhujee, about 60 years ago the river was arrested by an enormous avalanche of its banks, and during a period of 40 days not a rill of water escaped. This fact was related by a very old man with a hoary head who was eye-witness to the

phenomenon. It was brought to notice by observing a temple, or sanctuary, high above the level of the stream, eaten, as it were, away by the effect of water. The gatherings of such a stream, for 40 days must have been prodigious, and the consequences were duly contemplated by the inhabitants, for before the breach of the barrier a sort of telegraphic signal was established along the course of the river to announce the approaching event. The Raja's place at Belaspoor was swept away by the vast and overwhelming tide, and many lives were lost. An earthquake is supposed to have been the agent in the avalanche. The Sutlej at the point recorded to have been arrested is very narrow, and the cliffs on each side appear to be but a few yards apart.

76. During the winter months there is a ford across this river opposite Soonee, and near this, trans-Sutlej, are found hot springs which seem to deposit sulphur and nitre.

77. At Rampoor, the capital of Bussahir, the river is already at the height of 3,200 feet, agreeably to numerous barometrical observations taken at different seasons of the year, and is here (under the town) 210 feet broad, very rapid and obstructed by blocks of rock. The rise of the river from the last point of observation is about 31 feet per mile direct distance, but little more than half by the margin. A rope bridge, or jhoola, is thrown across it here. This consists of 5 or 6 thick grass ropes drawn as tight as they can be made, but necessarily forming a catenarian curve; from this a loop or cradle chair of rope is suspended, being attached to a piece of wood scooped or hollowed so as to traverse the ropes, and in this the passenger takes his seat and is drawn across the foaming torrent, at a gerat height above it and under considerable,

agitation. A little below this point, where the stream is clam, it is crossed by inflated skins, but the space is so narrow that accidents often occur, and the adventurer on his frail vessel becoming entangled in the current is overwhelmed and perishes. The course of the river hitherto has been from Loodiana to Belaspoor, within a few points of north-east; thence to Rampoor within a few points of east. Above this the rise of the level goes on very rapidly, vast masses of feldspar arrest the stream in many parts; but it often softens into deep blue calmness, and is hemmed in by mural cliffs; the course is now nearly east; the country assumes a ruder aspect: the mountains are bolder and rise to a vast hight; the scale of nature enlarges, and the Sutlej seeks its course with loud noise. Hitherto it has diminished scarcely anything, no feeder of any size occurring except the Naugree, which rises from the snow in the Moral ridge falling into the Sutlej four miles below Rampore. At Wangtoo, 41 miles above Rampore, we find the river rolling in a granite channel. At a height of 5,200 feet the stream is amazingly rapid, and the solid rocks of the banks, scoped and eaten away by incessant friction, record the violent action of the waters. The river is crossed here by a sanga, or wooden bridge, 82 feet within the margin. On the Gorkha invasion this bridge was demolished, and such is the inaction and frugal policy of the Government, that, until aided by British liberality, its renewal was never contemplated, a rope bridge sufficing for foot passengers, but very inconvenient and dangerous. Above this point the periodical rains are very light, the river being now environed by the Himalaya peaks. The stream is observed to be here fullest in May and June, or beginning of July, from the snow thawing under the powerful influence of the sun, after this period it receives very little accession from the rains, and the snow is then both very elevated and much diminished in bulk. In the interval between Rampore and Wangtoo the medium rise of the river is 74 feet per mile. Beyond this the stream is dreadfully agitated; falling over a great declivity and bounding like the swell of the sea, vast masses of rock insulated in it, and the borders lined by acres of blocks, it tears its way in form and whiteness overhung by impending avalanches, menaced by eternal snow.

Under the village of Brooang at the confluence of the Buspa, the head of the Sutlej is found at 6,000 feet; the direct distance being eight miles from Wangtoo gives the rate of the acclivity 100 feet per mile; the stream is here broad and slightly agitated : bed and borders of sand and pebbles. The Buspa is a principal feeder of the Sutlej and may contribute one-fifth to its size. It rises from the eternal snow in a ridge of mountains that is crossed by a pass to the Tartar village of Neelung on the banks of the Junnubee, or veritable branch of the Ganges.

78. The next point where the Sutlej has been correctly observed is at the village of Poaree, in the division of Kunawar named "Tookpa"; the road distance is about 11 miles by the bank of the stream, the direct about 7 along the margin of the river, which is much less ruffled than in its other reaches, till near Poaree, where it dashes amongst rocks and spreads out to nearly 100 yards. In July the body of water is immense; the whole expense of bed is overflowed and trees appear studded in the water, their tops only visible. There was formerly a sanga, or wooden bridge

here (Poaree), but it was destroyed on the approach of the Gorkhas, and while they were yet at Wangtoo a temporary rope bridge was a created whenever required. The breadth of the river and the sandy border are unfavourable for a good sanga. The level is here about 6,500 feet, having risen at the rate of 66½ feet per mile. In this neighbourhood the vine comes to great perfection. Thence on to Rispee, 13 miles by the road and about 8 direct distance. In this interval the river exhibits an intercepted line of whiteness, being choked by granite blocks and hampered by mural cliffs.

79. A little below Rispee the stream softens and spreads over a large expanse of sand and pebbles, and is intersected in the cold season by small islands : the surface is scarcely ruffled; the elevation is here 7,000 feet above sea level, the rate of a ascent 80 feet per mile. In this last space, or between Poaree and Rispee, is situate the Kylas cluster of peaks, all of granite, their feet washed by the Sutlej, which hourly carries away the lose fragments and mouldering portions of the rock, while the frosts are gradually crumbling down their elevated cliffs and levelling the summits, which are covered by eternal snow, and attain the height of 21,000 feet. They rise up very abruptly and spire into sharp crests (at one reach) of 14,000 feet perpendicularly above the Sutlej.

80. Beyond Rispee the rocks change to slate and gravel and crumble at their surface, and the Sutlej preserves a more tranquil character and is often seen from the heights which the road traverses, deep blue, and still at a vast depth below.

81. Proceeding towards Nissung, a Tarter village of Kunawur, the road rises to the

height of 13,700 feet, and the river is seen below under an angle of 40 degrees.

82. At Numptoo sanga, or wooden bridge, the Sutlej has an elevation of 8,200 feet, which by inferences from the level at Rispee, where observations are more numerous, gives the medium rise in the direct interval of 51 miles at the rate of 80 feet per mile. The river is here 80 feet broad, under the sanga, which is 80 feet above the water. The rapidity and violence of stream is here excessive, and in one point the whole volume of water is compressed within a space of about 10 feet. This spot is altogether very singular; the site of the bridge is upon the wrecks of an ancient avalanche on one side, and on the other rests upon a neck of rock eaten into by age and the action of the stream.

83. At the cradle bridge of twigs at Numgea, seven miles direct distance higher, the river is 8,600 feet and much reduced in size, the Lee, or Spittee river rising in a ridge of land, which on its northern slope gives accessions to the Indus through Ludauk, joining about a mile below and little inferior in bulk to the Sutlej itself. The contact of the rivers is singular. The Spittee issues from a dark worn ravine, in a tranquil stream, and its waters scarcely mingle with the Sutlej, till she is swallowed up in the clamour of her impetuous consort. The twig bridge under Numgea is both an insecure and tedious method of crossing, less safe, but more convenient than a jhoola. It is formed by a line of ropes made of twisted twigs which stretches from bank to bank; above this and on each side are leading ropes of the same materials, and connecting ribs at intervals of a few feet, forming altogether a trough, in which the passenger presses forward. The wind frequently gives the whole

a literal motion, very disagreeable to the feelings of the person crossing. This sort of bridge frequently gives way and people are lost.

84. Beyond Numgea all the way to Shipkee the river foams amongst rocks, with an acclivity of 150 ft. per mile, tearing its way in surf and noise; the distance by the road is nine miles and by the river's edge seven. The pass to the Chinese territory is at the height of 13,500 feet. The Sutlej is here confined within vast mountain masses, copped by perpetual snow. The loftiest is Purkwal, supposed to be 22,500 feet; a pile of granite, yet an appalling monument of the slow but certain ravages of time and weather, it appears daily to be falling to pieces; immense masses are hourly seen falling into the abyss.

Village of Shipkee

85. Shipkee is the frontier village of Chinese authority by the course of the Sutlej; it is a populous place and the inhabitants appear to live comfortably. The extreme elevation is 10,600 feet. Apricots flourish luxuriantly. There are no periodical rains here as in the plains of India, and in July and August the sun shines out its course, the sky being rarely obscured. The temperature at this time is between 53° and 73°. The crops are cut in August. The village lies on the slope of snowy mountains, facing the tableland. The Sutlej runs past it at the distance of a mile. It was visited by Dr. J. G. Gerard and observations of the barometer made to ascertain the height of the stream, which is 9,700 feet above the sea. A bridge of twigs crosses it here. Shipkee is on the high road to Garoo and Mansurowur. A post of horses communicates with the interior. The course of the river from Wangtoo to Shipkee is to the north of north-east, and hence to its source south-east.

*APPENDIX IX**

KINNAUR IN GLOSSARY OF TRIBES AND CASTES IN PUNJAB AND NWFP (1911)

KANAURI, KANAWARI,** an inhabitant of Kanaur, the valley which, lying on the Upper Sutlej, forms an appanage of the Bashahr State. Its inhabitants are almost entirely Kanets or Jāds, but differ as completely from the Kanets of Bashahr proper as do the Lāhula Kanets from those of Kullu.

Besides the Kanets or Jāds the only two castes in Kanaur are the Chāmang,*** who make shoes and weave, and the Domang,† who are blacksmiths and carpenters.

Water or cooked food which has been touched by the lower castes is not used by Kanets, nor are people of these castes allowed to enter a Kanet's house. If a Kanet eat such food inadvertently he applies to his Rājā who bids him make expiation (*prayaschitta*) and pay some *nazrāna* or forfeit. This custom is called *sajeran* or *sacheran*.

The Kanets of Kanaur are said to be divided into three grades, each comprising a number of septs, whose names do not appear among the Kanets of Bashahr proper.

The Kanet septs of Kanaur, according to their Geographical Distribution

1ST GRADE KANETS

Pargna Rajgaon

Bairyan	Moelas	Sakhān
Skāmtas	Sanā	Rokérū

* Extract from: Glossary of the Tribes and Castes of the Punjab and Northwest Frontier Province based on Census Report of 1883; by Sir Denzil Ibbeston; Vol. II A—K. Lahore 1911)

** Kanaura appears to be the more correct form.

*** The Chamārs of the Plains, doubtless.

† The Doms of the plains.

Wāngmo	Shwāl	Deran
Thāngar	Daprāṭo	Dalyan
Dāmes	Bitaryan	

Kanet Septs in Kanaur

Parganā-*Shuwa*

Bōres, Tib. Pores	Khādura	Lāspā
Ranshyān	Barji	Shilī
Dhanshyān	Shyālṭū	Gyōlang
Farakpā	Tholpā	Thārmi
Pālsar	Lokṭas	Puān, Tib. Puang
Alḍana	Pāngṭū	Makalā
Chhugpō	Shuryān	Mispon or Mishpon

Pargana-*Tukpā*

Bīsṭ	Rāṭhū	Dhangch
Kāllam	Nyokché	

Parganā *Outer Tukpā*

Repālṭū	Chāngkung	Dūdyān
Chethā	Fanyān	

Parganā *Pandarabīs*

Yulān	Chogla

Parganā *Thārābīs*

Tyūras	Jogṭo	Zinṭū

2ND GRADE KANET

Parganā *Inner Tukpā*

Brālbang	Mojrang	Shyāli
Chāmpo	Pānkar	Soṭhā
Kāṭhū	Rākshas	Ungyā

Parganā *Shūwā*

Kharyān	Turkyān	Khinpa, Tib. Khyimpa
Shyūna		

Parganā *Rajgaon*

Anchhān, Tib. Angchan Mashān . . .

3RD GRADE KANETS WHO WORK AS POTTERS

Wanghhāng	Wāzā	Méwar

Titles of Officials

1. *Chāres*, the hereditary headman of a village (in each village).
2. *Grokch*, the hereditary *kārdār* of the village deity, who speaks on his behalf.
3. *Māthas*, the hereditary *kārdār* of a deity. His duty is to petition the deity on behalf of the public.
4. *Pujyāres*, whose hereditary duty it is to worship the deity : Nos. 2, 3, and 4 are found in every village where there is a deity.
5. *Bathungrū* an official like the *dafādār* of the State.

In the Kanaur valley Buddhism is the dominant faith, but though the social customs of the people generally resemble those of the Hindus, the observances bear Tibetan names, and the ritual is conducted in that language.

BIRTH CUSTOMS

During pregnancy the following chant is sung : 'O goddess Tārā, I bow down to thee, be pleased to bestow on this woman thy choicest blessings'. And a charm written on a bit of paper or birch-tree bark is tied round the woman's neck.

On the birth of a son the Goddess Ḍolmā is adored, and the chant called *Bhum Chung*, which runs : *Om tāyāthā gāte gāte pārā gāte swāhās* ('May God bless the child') is sung. The old women of the family perform the midwife's functions; and for a fortnight the mother lives apart, being debarred from touching anything. At the end of that period she and all she possesses are sprinkled with cow's urine mixed with Ganges water, as among Hindus. The child's horoscope is cast by a *Lāmā*, who also names the child when it is 15 days old, or on any other auspicious day. It is generally brought out of the house for the first time at the full moon and, if possible, at an auspicious moment, when one or two months old. Charms for its long life are also made by the Lāmās.

A boy's head is shaved when one year old, the Lāmās performing a *hom*,* *pujā*, or *pāth*, sacrifice. As the Kanauris

* *Hom* is a rite in which flames are fed with clarified butter mixed with barley and sesamum seed; if possible almonds and dried grapes are also mixed in it. Pūja is an offering to the deity of a lamp fed with butter, water, flowers, sweetmeats fruit, etc., while *pāth* consists in reading or reciting the Tibetan scripture called Chhas or Chhoss.

only rank as Shudras, they are not entitled to wear the sacred thread, so they wear instead a *kaṇṭhi*,* or necklace from the age of 8.

MARRIAGE CUSTOMS

The marriage customs in Kanaur resemble those of the Tibetans. Brothers marry a joint wife, the Lāmās solemnizing the wedding by chanting certain hymns and worshipping the gods or goddesses, goats also being sacrificed.

The nuptial rites in Kanaur are peculiar. In the first place the amount of the *ḍherī* is unusually high, varying from Rs. 100 to Rs. 1,000.** The custom as to dower is also different. Many people give the bride as many pewter vessels as there are in the bridegroom's family, but ornaments, he-goats, cows, etc., are also given. The wedding is thus solemnized :

One of the brothers, most usually the one who is the bride's equal in age, goes with some of his relatives to her father's house on the day fixed by the *lāmā* (priest).*** There the party are well entertained, and the *lāmā* solemnizes the wedding by reciting some chants in Tibetan after the Tibetan manner. Next day they return to their own house with the bride richly dressed and adorned. On reaching home the bride is made welcome, especially by her mother-in-law. After a religious ceremony, the bride's right hand is held by all the bridegroom's brothers, and then all of them are deemed to have married her. A feast is then given to all who are present, and the *Lāmās* and musicians are fed. This marriage is a valid

* The *Kaṇṭhi* is an ordinary necklace made of *tulsi*, the holy basil (Ocymum sacrum). These *kaṇṭhis* are generally made in Hardwar, Bindraban, Ajudhya, and Benares.

** The *ḍheri* prevents a woman's going to another man, as only he who takes on himself the responsibility for it is entitled to keep the woman. It is a sum paid to the bride's guardian by those of the bridegroom, and must be refunded to the latter if the marriage turns out badly, e.g., if the wife leave her husband and go off with another man, he was to refund the amount to them.

*** The wife is often older than her husbands, or than some of them, and her age, especially if coupled with a sharp tongue, gives her a decisive voice in the family councils.

one. The child of an unmarried girl is called *puglang* (bastard), and has no right to anything by way of inheritance. Such children live by service and marry with some one of their own class, i.e., with a *puglang* or *puglakch*.

In case all the brothers have only one joint wife, there can be no question as to the right of inheritance. And just as the bride's mother-in-law is mistress of the family, so on her death the wife succeeds as its mistress. Thus the movable and immovable property of a family remains in its joint possession and is never divided. But the custom of polyandry is now dying out by degrees.

DEATH CUSTOMS

As the trade and wealth of Kanaur increase and its people come more in contact with India, they are rapidly abandoning the old customs, such as *ḍūbant* (drowning), *phukant* (burning), *bhakhant* (eating), etc. This last method of disposing of a dead body was formerly observed only by the inhabitants of *Hangrang ghorī* who are called by the Kanaur is Nyām, and by the Kochī or Pahāṛi people, Zāṛ or Jāṛ, Zāḍ or Jāḍ.

The *lāmās* used to consult their scriptures and advise as to the disposal of the dead according to the time, etc., of the death, but now the Hindu *shrādhs*, and so on, are observed. The only old customs which survives is the annual *shrādh* called *phulaich**, in which a he-goat reared in the dead man's name, is dressed in his clothes, sacrificed and eaten by the members of his kindred.

At a death-bed, grain is distributed among all those present, and the *lāmās* read from Buddhist writings. The body is burnt on the same day, or at latest on the next. Drums, *sanāīs*, *karnāls*** and conches are played when the corpse is carried to

* Fulaich or Phulaich, from Hindi *phūl*, flower, is so called because Kanauris do not wear new clothes till one year after a death in the family, but after performing the *dujang* they may wear flowers and new clothes.

** The *sanāī* and *karnāl* are both musical instruments used in the hills. The former is made of wood and is about a foot long with seven holes on which the fingers are placed while playing and its sound is like that of an *algoja*; the latter is made of brass and is like a long horn with a round, broad mouth; in sound it resembles the conch.

the burning-ground. Some of the bones are picked up, and sent either to Mānasarowar in Tibet, to Rawālsar in Mandi State, or to the Ganges.* In the deceased's room a lamp is kept burning for seven days from the death, and incense is also burnt in it. The *chholpa* (Hind, *kiria karm*) is performed from the eighth to the tenth day; all the deceased's clothes are given to the *lāmās*, with other gifts. The *panchaka* or group of five constellations** is inauspicious for the family of one who dies under it, and to avert the evil, images of roasted flour are made and burnt with the corpse, to the accompaniment of Tibetan chants.

After 15 days the *lāmā* does a *hom pūjā*, and *pāṭh* reciting Tibetan chants of purification. This ends the period of mourning. After a year the *phulaich****. is observed, by giving food and clothes to a *lāmā* in the deceased's name; and until this is observed the family must not wear any new clothes, etc. The *shradh*, called *dujang* in Kanauri, is also solemnized by the *lāmā*. The burning-grounds are haunted by Mashān, Rākshāsa, Shyūnā, and Khar-shyūnā, of whom the first two are conceived of as evil spirits or demons and the two latter as Jack-o'-lanterns or ghosts.†

The following chant is repeated by the *lāmā* more than a thousand times to exorcise an evil spirit from a man or woman : *Om Bājrā kīlā kīlāyā dimo shākchā uchā thāyālā fāṭ.* Any one bitten by a mad dog is healed by repeating the following chant a thousand times : *Om khu-khu rāchā khā-thām dewā chāng ghī dwīshok.*

* Taking the bones to the Ganges is said to be quite a recent innovation-only dating back two or three years.

** The five nakshatrās are Dhānisṭha, *Shatbhishā, Pūrvābhadrāpādā, Uttarābhādrāpādā* and *Rewati.*

*** Phulāich is also the name of a fair held in October every year at Brāling, near Ropā. See Ukhyāng in the List of fairs.

† Mashān and Rākshasā are of course Sanskrit terms. The other two are Kanāwari, possibly corruptions of Tibetan words. It is worth remarking that Mashān Shyūnā, and Rākshasā are also septs of Kanets' found in Mellam, Āsrang, and Rirang villages respectively.

A CHRONOLOGICAL LIST OF THE BUDDHIST RELIGIOUS OBSERVANCES IN KANAUR

1. The Kāngso, a religious ceremony, in which the *hom pūjā*, and *pāṭh* are performed by the *lāmās* and *zomos*, observed in every temple throughout Kanaur on the 8th, 10th, 12th, and 14th of the bright half, as well as on the full moon and *amāwas* of each month.

2. The Zinshok, celebrated in Kānam village on the 8th of the bright half, as well as on the full moon of each month, including the *amāwas*.

3. The Torgyā, performed at Kānam, once on the 14th of the bright half and again on the full moon of Phāgun.

4. The Tona, also celebrated at Kānam on the 11th of Chait for one day.

5. The Tībāngmā, performed at Kānam on the 20th of Paūsh.

6. The Kutimf, also celebrated at Kānam on the 13th of Phāgun.

7. The Nāmgang, also observed at Kānam for two days from the *amāwas* of every month. *Hom pūjā*, and *pāṭh* are performed by the *lāmās* and *zomos*.

8. The Shibrāt (Sanskr. Shivarātri, the birthday of Shiv or *Mahādeo*), is a religious ceremony not only of the Hindūs but also of the Buddhists. It takes place on the 14th of the dark half of Phāgun, on which day the people adore Shib, whom they call Lofan, and distribute food among relatives and friends.

9. The Shopṇeṭang, (Sanskr. Shrāvanārchana, meaning 'worship of Sāwan'), is celebrated at Grāmang or Kaṭhgāoṇ in Bhābā *pārganā* on the full moon of Sāwan. About a dozen young men, taking with them cooked food for three days, go out to gather wild flowers and plants from the loftiest snow peaks. They pass two nights there, collecting various kinds of wild flowers and plants, and on their return they are received with joyous music by the villagers. The garland which they bring from the forest is offered to the deity, and they then, together with women, dance and sing songs.

10. The Lāmā-pāzā, a Buddhist religious rite, is observed at Lābrang, Shūwā *parganā*, on the *amāwas* of Chait. The *lāmās* and *zomos* devote themselves to the worship of the deity Chhakoling Dambar, while dancing and singing are performed by men and women with great rejoicings.

11. The Jāgro (Sanskr. Jāgarana, a vigil), is also a religious ceremony, observed throughout Kanaur on the 20th of Bhādoṇ. The night is spent in singing and dancing to music, and worship of the deity is performed in all the temples.

A List of the Fairs Held in Kanaur with a Brief Description of Each

1. Losar, or New Year's Day, is observed at Kânâm for three days, from Paush *shudī* 13th till the full moon of Paush. All the people assemble to ask the *lāmās* about their gains and losses during the coming year. It is the most characteristic fair of Kanaur. Feasts are given to friends and relatives, but dancing by men and women to music is the chief function.

2. The Kāngyūr-zālmo (fr. *kāngyur*, library, and *zālmo*, a visit), takes place on the 15th *gatē* of *Hāṛ* (Ashāṛh) at *Kānam*. People visit the Tibetan Library, called Kāngyur-tāngyur, in the large village of Kānam.

3. The Menthako fair also takes place at Kānam on the 20th *gatē* of Bhadon (August) and lasts two days. The chief event at this fair is a pony-race, feasting, drinking, dancing to music and singing.

4. The Khwākchā fair takes place at Kānam and lasts for 5 days from the 20th *gatē* of Māgh, ending on the 25th. The nights are passed in dancing and singing to music before the temple of the deity called Dābla.

5. The Gāngā fair takes place in Chāngmang forest above Lippā, in Shūwā *parganā*, on the full moon Kātik. Men, women, and children climb to the Chāngmang forest, and eating, drinking, dancing, and singing are features of the festival.

6. The Jokhyā-kushimig and Jokhyā-chhugshimig at Kānam are important festivals, at which visits are paid to relatives and friends, on the 13th and 14th *gatē* of Māgh (January).

7. The Ukhyāng (fr. *ū*, a flower, and *khyāng*, a sight of) is the most remarkable fair in Kanaur. The people go to

the high ranges to gather wild flowers and leaves, and offer a large garland of them to the deity. Men and women in rich attire also dance and sing a song which is roughly translated thus :

> "The fair called Ukhyāng is held first at Rupī village* in honour of the village deity named Tèras**, on the 10th of Bhādo.
> In Bārang village*** it takes place on the 20th *gatē* of Bhādo, when the upper forests are full of wild flowers and plants.
> For whose sake is this monkish garland† O Nāges†† of Bārang, its for your good sake.
> The Ukhyāng fair takes place when the forest is dry, in the dry forest there are no flowers.
> What is to be done then? Again we say what is to be done?
> Behold a garland of *rāchū* kānang;††† to whom we are to offer it?
> It must be offered to Mārkāling.@
> Again to whom should we offer a garland of Shishyur?@@
> We must offer it to the deity of Yānā or Mellam, by name Gāndrapas.@@@
> Where is the remainder of the fair held? The fair of Maheshras,× the Bhābā *parganā*×× deity, takes place when the autumn moon is full.
> A handsome Rājā is Rājā Shumsher Singh
> And Maheshras, the deity of Bhābā.

* Rupi is a village in Pandarabīs *parganā*
** Teras, the deity of Rupī village
*** Barang, a village in Inner Tukpa *parganā*
† Loshgar, the monkhood flower
†† Nāges, the deity of Bārang village
††† Rāchū-kānang, a plant which has leaves like a calf's ears whence its name.
@ Mārkāling, deity of Khwāngī, a village in Shūwa *parganā*.
@@ Shishyur, a plant found on the snowy peaks.
@@@ Gāndrāpās, the deity of Yānā or Mellam, a village in Rajgāon *parganā*.
× Meheshras (2nd), the deity of Bhābā *pārganā*.
×× 8. Bhābā is a *parganā* in the Wāng valley.

Like Shūwa Chandika* is beautiful
In Tukpā *parganā* there are nine water channels,
But Shūwas *parganā* has only one."

8. The Shogch fair is held at Chīnī and lasts for 5 days, from Mangar *shudi* 10th to the full moon of that month. People from all the surrounding villages assemble to dance and sing and a great deal of merriment results.

9. The Rāthin fair is also held in Chīnī on the 1st of Paush and is celebrated by dancing and singing.

10. The Agtarang fair at Richpā or Rispā in Inner Tukpā lasts for one day. All the people of the surrounding villages assemble, and dancing and singing before the temple of Kulyo deity are the features of the fair.

11. The Māng fair is also observed at Richpā and lasts for about a week from the 18th of Māgh. The *lāmās* and *zomas* devote themselves to the worship of Buddha, men and women dance and sing to music with great merriment till the end of the fair.

12. The Yungnas or Jungnas fair is also held at Richpā in Paush, the exact day being fixed by the *zamīndārs* to suit their own convenience, and it lasts for five days. Worship of Buddha is observed with general rejoicings. Eating, drinking, dancing, and singing are the principal features of the fair.

13. The Sherkan fair is held in Kānam on the 3rd of Kartik and lasts but one day.

14. The Dumgyur-zālmo fair takes place at Kwāḷdā, in Shūwā *pārganā* on an auspicious day appointed by the zamīndārs in Hāṛ (Āshārh). *Dumgyur* means a Buddhist praying wheel, and *zālmo*, a visit. The people visit the huge praying wheel, and turn it round to the right as often as they are allowed.

15. The Kailās-zālmo, or 'the visit to the Kailās mountain', is celebrated at Pīlo or Spīlo, in Shūwā *parganā*, on any auspicious day in Hāṛ fixed at the will of the *zamīndars*, and lasts one day. Worship of the Kāilās mountain is performed

*Shūwāng Chandikā, the goddess of Koṭhi or Kosṭampi, a village in Shūwā *parganā*.

with great rejoicings, dancing and singing, being the main features of the fair.

16. The Khepā fair is observed throughout Kanaur, for three days, from Maṇgar *badī saptami* to Maṇgar *badī dasmī*. The people bring thorns and put them on the doors of their houses, in order that no evil spirit may enter and on the 3rd day they take all the thorns outside the village and burn them, as if they were burning an evil spirit. Dancing and singing with music are main features of the fair.

17. The Rās-kāyāng (*rās* Sanskr. *rāshi*, a zodiacal sign and *kāyang*, Sanskr. *kāya*, body) is the day on which the sun reaches the zodiacal sign of Aries. In India known as the Mēshi-saṇkrānti or Vishuvasaṇkrānti, throughout the Simla Hills it is called Bishū. This fair is celebrated throughout Kanaur and the Simla Hills on the 1st of Baisākh. The houses are well whitewashed and decorated, and dancing and singing with great rejoicings are its main features.

18. The Lābrāng-zālmo fair takes place at Kānam on the 17th of Jēth. At this fair people visit the temple of Dāblā, and dance and sing there with great rejoicings.

19. The Chhokten-zālmo fair is held at Lābrang, in Shūwā *parganā* on the 15th of Hāṛ. People visit the temple called Chhokten at Lābrang. Singing and dancing to music are its main features.

20. The Suskar fair is observed in Kothī or Kostampī as well as elsewhere, about a week from the 9th of Phāgun. Two parties, one of young men and the other of young women, fight with snow-balls until they are tired. Singing and dancing to music before the goddess Shūwāng Chaṇḍikā are the main features of the fair.

21. The Jangang fair also takes place in Koṭhī on the 3rd of Māgh, and lasts for a day. Dancing and singing songs to music, and worship of the deity are performed with great rejoicings. *Jagang*, from Sanskr., *Yajna*, means sacrifice.

22. The Bishū fair is the same as the Rās-kāyāng, which takes place on the 1st of Baisākh. In Upper Kanaur the people call it Rās-kāyāng, and in Lower Kanaur, Bishū.

23. The Bāng-kāyāng fair is held at Grāmāng or Kaṭhgāon, in Bhābā *parganā*, on the full moon of Paush. All the Bhābā

people assemble in the temple of Maheshras and worship him. Dancing and singing are the main features of the fair.

Monasticism

Kanet girls, who do not marry, but devote their time to the study of the Tibetan scriptures are called *zomos* or *jamos*. They live in nunneries. The two principal nunneries are at Kānam and Sunnam, and in these a great number of *zomos* live. Besides this, every village has a few *zomos*.

Kanet boys, who learn the Tibetan scriptures, and are well versed in the Buddhist doctrines, are called *lāmās*. They live in monasteries and are looked upon as very holy. In fact they are the priests of all the Kanets. There are several monasteries of these *lāmās* in Kānam, Sunnam and other villages. *Lāmās* are either Gyolang or celibate, like the Brahmachārī, or Ḍugpū, who marry but never shave the head.

The *lāmā* is consulted with regard to every important undertaking. Thus he is asked to name an auspicious day for beginning to plough or sow, and at the time ascertained he recites chants like the one beginning : *Om akānī nikānī ambitā mandātē mantālē swāhā*, 'May the gods bestow on us abundance of grain'. When a new roof is put on a temple, which is called *shānt** the *lāmās* perform a ceremony, reciting charms and performing *hom*, with the sacrifice of sheep and goats. This is called *paresṭāng* (Sanskr. *pratisṭhā*, consecration). When a new house is ready the *lāmā* fixes the time auspicious for its occupation, and the owner, dressed in new clothes, is then taken into it with his wife, who rings a bell. This is called *gorāsang*.** New grain is first offered to the village-god and may then be eaten.

*From Sanskr. *shānti*, peace.

**From Sanskr. *grihapravtishṭha*, entering in a house : it is called *gharasnī* in the Simla Hills.

CULTS IN KANAUR

AN ALPHABETICAL LIST OF THE DEOTAS IN KANAUR, TOGETHER WITH THE NAME OF THE VILLAGE IN WHICH EACH IS LOCATED

1. Bādrināth, at Kāmrū or Mone village.
2. Bhīmākalī, at Kāmrū or Mone village (Also at Sarāhan).
3. Chhakoling Dāmbar, at Lābrang village in *parganā* Shūwa.
4. Chāndikā, at Ropa village in Shūwa *pargana*, Gangyul *ghori*. Also at Yāwring village, Shuwā *pargana*.
5. Ghhwedung, at Chāngo village in Shūwā *parganā*.
6. Dāblā, at Kānam, Dābling, Dūbling, Lío, Spūwā or Poo, Shyashō, in Upper Kanaur.
7. Deodūm, at Kako, in Shūwa *parganā*.
8. Gyāngmāgyum, at Jāngí, in Shuwā *parganā*.
9. Kasūrajās, at Rirang or Ribbā, in Inner Tukpā *parganā*.
10. Khormo, at Pīlo or Spílo, in Shūwa *parganā*.
11. Kulyo, at Richpā or Rispā, in Inner Tukpā *parganā*.
12. Mahēshras, at Shungrā or Grosnam in Thārābīs *parganā*, at Grāmāng or Kathgāon, in Bhābā *parganā*, and at Chūgaōn or Tholāng in Rajgāon *parganā*.
13. Mārkāling, at Khwāngi in Shūwa *parganā*.
14. Māthi, at Chhitkūl, in Outer Tukpa *parganā*.
15. Milākyūm, at Akpā village in Shūwa *parganā*.
16. Nāges, at Bārang, Brūāng.* Chāsāng, Chhotā Kāmbā, Kilbā, Mewar, Mírū, Sānglā, Sāpní or Rāpāng villages.
17. Nāgin, at Bari village in Thārabis *parganā*.
18. Nārāyān, at Barsering village in Outer Tukpā *parganā*.
19. Narenas, at Asrang Chīnī, Shohāng, Urnī, and Yulā Villages; and also at Chūgaon, Grāmāng and Shungrā, with the three Maheshras.
20. Orming, at Morang or Ginam village in Inner Tukpā *parganā*.

* Nāges deotā in Sānglā is thus addressed by the *pujyāres* in worship :— O thou, who livest within the wall, who livest in holes, who canst go into a vessel, who canst swiftly run, who livest in the water, on the precipice, upon the trees, in the waste land, among the meadows, who hast power like the thunderbolt, who livest within the hollow strees, among the rocks, within the caves, be victorious.

21. Pāthoro, at Rārāng and Punnam villages. Shūwā and Rajgāon *parganās*.
22. Rogshū, at Rogī village in Shūwā *parganā*.
23. Shankras, at Pwārī or Por village in Inner Tukpa *parganā*.
24. Shanshras, at Rākchham village in Outer Tūkpā *parganā*.
25. Shēshēring, at Pāngi village in Shūwā *parganā*.
26. Rāpukch, at Thāngi village in Inner Tukpā *parganā*.
27. Shūwāng Chandikā, at Kosṭampī or Koṭhī village in Shūwā *parganā*.
28. Tārāsāng, at Trāndā village in Thārābis *parganā*.
29. Tērās at Rupī village in Pandrābis *parganā*.
30. Tungma Dāmbār, at Gyabung village in Shūwa *parganā*.
31. Ukhā, at Nachār and Bārā Kāmbā villages, Thārābis and Pandrabis *parganās*.
32. Yulshā, at Sunnam village in Shūwa *parganā*.

APPENDIX X

GLOSSARY

Aanch	: He-goat
Aar	: Handle of a plough
Aari	: Saw
Abadi	: A populated spot e.g. village
Achar	: Pickle
Adna malkiat	: Inferior proprietary rights
Aga	: Fire
Akhara	: Wrestling arena
Akrot	: Walnut
Almalik	: Superior landowners
Alu	: Potato
Ang	: Cave
Anguri	: Grape wine
Apsara	: Nymph
Arbo	: A vessel for washing hands
Arri	: Handle of a plough
Arz-i-arsal	: A memo given by the *patwari* to the village headman indicating the distribution under the proper heads of the money to be taken to the tahsil
Ashram	: Hermitage
Ashtdhatu	: Compound of eight metals
Atal	: Soap-nut
Atta	: Flour
Auras	: Carpenter
Awal	: First class
Bachh	: Distribution and imposition of the state demand of small kind
Bagicha	: Area under fruit trees
Bagra	: Inferior
Baimi	: Peach

Bajra	: Millet
Bakhal	: Unirrigated land
Bakhor	: She-goat
Balang	: Cow
Balwadi	: Nursery school
Ban	: A musical instrument
Bando bast waziran	: Settlement conducted by waziers
Bandrus	: Monkey
Bangparshimig	: A local dance
Baniya	: Shopkeeper
Baramda	: Verandah
Barani	: Unirrigated
Bash	: Goat-hair rope
Batai	: Exchange
Bathu	: A foodgrain
Batich	: A cup
Beeyang	: Seed
Belding Sarmu	: A marriage ceremony
Bennang hachis	: A form of love marriage
Bergi	: A woollen shawl
Berka	: Stick for threshing
Beru	: Scheduled-caste
Bespa	: Traveller
Bethu	: Labourer, menial worker
Bham	: A large kettle-drum
Bhandari	: Treasurer
Bekhal	: A shrub
Bhiu	: Calf
Bhojpatra	: Birch
Bhut	: Evil spirit
Bijli	: Electricity
Bijul	: Lightning
Bila-sift	: Unclassified
Bilcha	: Pick-axe
Bitho pono	: Gifts allotted to the bride by the parents of bridegroom
Biting	: Wall

Biyoshimig	: A ceremony when the bride bids farewell to the party which accompanied her to the groom's house
Bogwan	: God
Bonyungchu-chashimig	: A local dance performed by menfolk only
Boong	: Donkey
Bospa	: Ash
Bot	: Skin
Bowli	: A well
Brahmachari	: Celibate
Bras	: A foodgrain
Breeme	: Female yak
Breja	: A local grain measure
Bugjal	: Cymbal
Bugo	: Coriander seed
Balzanuspono	: Woollen footwear
Burni	: An ornament given by the parents of bridegroom to the bride's parents on the occasion of engagement.
Buttoonggee	: State official
Cha	: Tea
Chadil	: Tea imported from Kali mpong
Chahar	: Leveller
Chaks	: Revenue estates
Chalang	: Dance movements
Chambaka	: A flower
Chambya Kukri	: Cock
Chamu baskat	: Woollen waist-coat
Chamu Kurti	: Woollen shirt
Chamu Sutan	: Woollen Trousers
Chano	: Hem
Chapati	: Loaf
Char	: Watchman
Chares	: A non-scheduled caste head functionary
Charnamrat	: Holy water
Charpaie	: Cot
Charsala	: Quadrennial recording of rights of land-owners/tenants

Charuk	: Wooden share
Chashionik	: Dance
Chatam	: Small sickle
Chathuk	: Iron share
Chebo	: Conical basket
Chet	: Marks on the lever
Chhachach	: Girl
Chhakcha	: Dry latrines
Chhalia	: Maize
Chhang	: Child
Chhanli	: Shawl
Chhatkang	: A house temple
Chhaunri	: Flywhisk
Chhemer	: Lizard
Chhesmi	: Woman
Chholpa	: Death ceremony
Chooma	: A name of dish
Chhosilma	: A death rite
Chhuba	: A garment
Chhudak	: Goat-hair rope
Chhwa phasur	: Alcohol distilled from foodgrains
Chigu	: A goat
Chini	: Sugar
Chironing	: History of the deity in verse
Choli	: Blouse
Chon	: A highway ghost
Chonegahba	: Five pieces of cloth
Chongmee	: Mane
Chopramgmar	: Butter
Chosten	: Buddhist shrine
Craw	: Hair
Dab-dab	: Marriage by capture
Dakhangu-phasur	: Grape win
Dakhung	: Grapes
Dam tangshis	: A form of love marriage
Dama	: Ox
Dandi	: Indigenous mode of conveyance (In Spilo village also used as bier)
Darohi	: Oath
Darosh	: Marriage by capture

Daru	: Gunpowder
Dashongi	: Head man of a pargana
Desang	: Village
Devi	: Goddess
Dhal bachh	: Aportionment of cash revenue
Dhan	: Paddy
Dhol	: Drum
Dhumang	: Smoke
Digra	: Silver hook
Doam	: Second class
Dogri	: An alternative farm hamlet
Dohri	: Silver hook
Dokan	: Shop
Domong	: Blacksmith
Dompo	: Leveller
Donktens	: Buddhist Shrine
Dukhrach	: Boy
Dum	: An organisation for regulating affairs of the village. Peaceful picketing
Dumko	: Large vessel used for distillation
Dung	: Conch
Dursa	: Pyre
Dusrang	: A hole in the roof for letting out smoke admitting in the light
Dyorang	: Temple (Hindu)
Dwarang	: Door
Gachhang	: A long strip of cloth tied round the waist
Galyoung	: Abuse
Gaonsar begar	: Village-wise forced labour
Gate-boba	: Younger father in a polyandrous family
Geethung	: Song
Ghani	: Bullock-driven oil crusher
Gharat	: Water-mill
Ghartangna	: Rent for water-mills
Ghasni	: Grass preserves
Ghoont	: Hill pony
Ghori	: Sub-division of a pargana

Ghoye	: A small woollen shawl
Ghrasni	: A ceremony performed on entering a new house
Girdawari	: Crop inspection
Goldes	: Vulture
Goling	: Hoe
Goom	: Bow
Gor	: Earthen vessel
Gorasang	: A ceremony performed on completion of the construction of a new house
Gori	: Coconut
Gotang	: Water-mill
Gram	: Village
Gram Sabha	: Village council
Gram Sewak	: Village level worker (male)
Gram Sewika	: Village level worker (female)
Grihapravesha	: A ceremony performed on entering a new house
Grokch	: Mouthpiece of a deity
Gudma	: Woollen blanket
Guthung	: Trough
Gyama	: Harrow
Hales	: Scheduled caste labourer
Halmandi	: A scheduled caste head functionary
Har	: Enticing away someone's wife
Heera	: Dimond
Hodd	: A dish
Hom	: Worship
Home	: Bear
Hosgang	: Waterfall
Hulas	: Ram
Id	: One
Izzat	: Self-respect or prestige. In Kinnaur the word is used in the context of the husband who is recompensed in cash by the seducer of the wife thereby the husband's izzat is supposed to be vindicated

Jalebi	: A sweet (honey comb)
Janekang or *Janetang*	: A form of marriage
Jamabandi	: Quinquennial recordings of the rights of land-lords and tenants
Jangmuch	: Mushroom
Jhula	: Rope bridge
Jinsi	: A ritual
Jinsrekch	: Sacrificial fire beside the pyre
Jomo	: Nun
Joo	: Cloud
Ka	: Walcut
Kag	: Crow
Kaib	: Needle
Kalu	: Scissors
Kamiz	: Shirt
Kanchong	: The right of the youngest in the ancestral property
Kanda	: Highland, away from homestead
Kankani	: A painted gate
Kangynr	: A bunch of a number of Buddhist scriptures and classics stored away in a temple is called *kangyur*
Kaning	: Stone mortar
Kanthi	: Necklace (invariably of beads)
Kapshaw	: Shoes
Karnahads	: Cesses
Karali	: Newly reclaimed land
Kargyul	: Cup
Kasang	: Bronze
Kayang	: A local dance performed in a circle
Kerche	: Lentil
Khal	: Oil cake
Khalangna	: Rent of oil-mill
Khanaur	: Chestnut
Kharif	: General term for crop harvested in summer
Kharshuna	: Light which is emitted by certain insects and plants or by phosphorescent fungus which glows like lamps

in the Himalayan hills. The local people say the lamp of the fairies.

Khashia : Non-scheduled caste
Khatak : A piece of cloth
Khatuch : Bridegroom
Khera : Mat of goat's hair
Kherach : A small mat of goat's hair
Kherang : Milk
Khindup : A form of marriage
Khlug : Cow
Khochar : Mule
Kholang : Thrashing floor
Khora : Cess in kind of *gur*
Khorich : Crucible
Khuchis : Marriage by capture
Khukri : Dagger
Khul : Goat and sheep skin
Khunds : Parganas
Khunkch : An evil spirit
Khura : A dish
Khurang : Cattle-shed
Khus : Sheep
Khyot : Spoon
Kiar : Irrigated land
Kim : House
Kimshu : Family god
Kod : A local grain measure
Kodu : Pumpkin
Kohlu : Oil-mill
Kote : Wooden chests
Koting : Basket
Kug : Owl
Kuhl : Water channel
Kuie : Dog
Kum : Pillow
Kunda : An idol
Kurnang : Wood carving
Kuwang : Tank
Kyalmung : Deodar

Kyo-rang	: Horse
Laan	: Air
Labrang	: Buddhist temple
Lagan	: Revenue
Lagang	: Buddhist temple of Mahadeo
Lageyach	: Rain
Lamthu	: A huge vessel made of brass or copper
Lapcha	: A square pile of stones considered holy
Leelaan	: Wind
Lhosnang	: Garlic
Lim	: Kail
Losoma Shalkid	: Thank you for wishing
Losoma tashi	: Wish a happy new year
Lotri	: Lotah
Lu	: Soft wool, lamb
Luchi	: A dish
Machhes	: Fish
Majomi	: Go-between
Mandi	: Market
Mane	: Buddhist shrine
Mangsho Tangsho	: Hide and seek game
Mant	: Female
Mant Kukri	: Hen
Mashan	: An evil being
Matung	: Earth
Matyang chilam	: Improvised smoke device
Matyang-Telang	: Kerosene oil
Meling	: Hearth
Melup	: Meteor
Menthok	: Ashes
Mey	: Fire
Mig	: Eye
Migpo	: Braw
Misa haqiyat	: Standing record
Mo	: Arrow
Mograng	: Wood carving
Muafis	: Revenue-fee grants of land

Mukhang	: A mask embossed and engraved to simulate human features with jewelled eyes
Mukhiya	: Village headman
Mul	: Silver
Mush	: Mash
Naar	: Wife
Nagus	: Spring
Nala	: A hill torrent
Namukkamal	: Incomplete
Nangch	: A small dish
Nango	: Platters
Naryal	: Big sickle
Nashing	: Yoke
Nautor	: New land brought under cultivation
Nish	: Two
Niza	: Twenty
Nyar	: Peas
Nyaya	: Dispense justice
Oo gobhi	: Cauliflower
Pa	: Four
Pag	: Turban
Paimaish	: To effect land measurement
Paithang	: Pulse
Panch	: Member of a judicial panchayat
Panchaka	: Five days in a month usually regarded inauspicious
Pancharattan	: Five precious things or gems
Panchet	: Panchayat
Pashm	: Wool of superfine quality
Path	: Reading from the scripture
Pathrang	: Leaf
Peetol	: Brass
Phasur	: Alcohol distilled from fruits or cereals
Phat	: Sack
Pho	: Deer
Phoch	: Donkey
Phorua	: Spade

Piada	: Messenger
Pinti	: Cess in kind of ghee
Pipli	: Chillies
Pishi	: Cat
Piyach	: Rat
Po	: Foundation
Pole	: Fried loaf
Poltu	: Illegitimate child
Pom	: Snow
Pono	: Footwear
Poreshing	: Lever
Porestang	: Roofing ceremony of a temple
Porsad	: Pudding
Pothee	: Holy book
Pu	: Nut
Pud	: Bhojpatra
Puglang	: Child of an unmarried girl
Pulashon	: A local dance
Purdah	: Veil
Pya	: Bird
Rabi	: Agricultural season of winter months
Rad	: Foodgrain
Ragee	: Mill-made cloth shirt
Rakh	: Country spirit
Rako	: Scorpion
Rakshas	: An evil being
Ral	: Rice
Rama	: She-goat
Rang	: Mountain peak, a horse
Razinama	: Agreement to the satisfaction of both parties. (In Kinnaur mostly pertains to the settlement of divorce)
Ree	: Neoza
Ri	: Edible pine
Rig	: Louse
Rim	: Field
Rinchot	: Gift given by the bridegroom to the bride
Robho	: He-goat

Robkar : Court's order
Roch : Musk Deer
Ron : Iron
Ronsheeng : Hornpipe
Ropya : Pigeon
Rote : Loaf
Rug : Stone
Ruk : Wine
Ruth : Chariot
Ruza : Old
Sahukar : Money-lender
Sajo githang : Farewell song sung exclusively on the occasion of the Sajo mela
Sambat tentis : Thirty-third year of the Bikrami era
Sambo : Laddle
Sammelan : Conference
Samundarang : River
Santhang : Temple courtyard
Sapis : Snake
Sapkhul : Skin bellower
Sarhaddi : Bordering
Sarpanch : Head of the judicial panchayat
Saru : Hailstorm
Satpanthang : A room in a *Kothi* of the deity
Semchain : Animal
Sening kunnu : Ceremony on the occasion of marriage when the bride or the bridegroom are introduced to their respective relatives
Shagar : Turnip
Shajra kishtwar : Field map
Shajra parcha : Patwari's copy of the field map
Shalwar : Loose trousers worn by women
Shamlat : Common land
Shanang : Lock
Shandrang : Pole fixed to the plough
Shaptuk : A death rite
Shapug : Flea

Shatardru	: Hundreds of currents (The Sutlej river)
Shaw	: Meat
Sheeng	: Wood
Sheethung	: Wood packer
Shew	: Cap
Shil	: Mid-day meal
Shing-tak-shiming	: A ceremony for the dissolution of the marriage
Shkumpatu	: Pitcher
Shonnal	: Hautboy
Shorcho	: Wooden Comb
Shoster	: Weapon
Sh[illegible]w	: Soil
Shradh	: Death anniversary
Shudung	: A kind of brewed liquor
Shu kothi	: A structure of a village deity
Shum	: Three
Shupakchu-khau	: Dinner
Siana	: Village elder man
Skyo	: Male
Soja	: Superior
Solak	: Road
Solch	: Plant
Sor	: City
Sora	: Small sickle
Sorgang	: Sky
Sorang	: Tank
Spal	: A local root
Spona	: Shoes
Sqalch	: Rack
Srien	: Tailor
Stish	: Seven
Tag	: Barley
Talung	: Key
Tapobhoomi	: Land of devotion
Tapru-se-balzanupono	: Footwear of an exquisite design
Tapru-se-chhanli	: A shawl with exquisite design

Tapru-se-dhori	: A local garment for women having an exquisite design
Tapru-se-sutan	: Trousers with exquisite design
Tarmim bandobast	: Revised settlement
Tee	: Water
Tee-palockt	: Frog
Teg boba	: Elder father in a polyandrous family
Tegsowang	: A local dance
Telung	: Oil
Thai	: A local measure
Thana	: A police station
Thatha	: A ceremony performed to ward off the evil spirit
Thepang	: Woollen cap
Thungru	: A local dance
Thur	: Tiger
Tockche	: Hand hoe
Togang	: Balcony
Toknya	: Menial worker
Toshimig	: To sit together
Tromung	: Copper
Troval	: Sword
Tug	: Six
Tupuk	: Gun
Umbu	: Wool
Van	: Steam
Vanchuling	: Earthquake
Vom	: Path
Wajib-ul-araz	: Village administration papers
Wass	: Honey
Wassphasur	: Alcohol distilled from honey
Wasyang	: Bee
Wizarat	: Area under the jurisdiction of a : wazir
Yangluk	: Woollen shawl
Yug	: Yak
Zalore	: Wooden fringe
Zama	: An earthen pot
Zira	: Cummin

Zo : Cross breed from yak and cow
Zod : Wheat
Zomba : Woollen show
Zongfa : A tray for winnowing
Zukhrang : Bush
Zung : Gold

Bibliography

PRIMARY SOURCES

A. Manuscripts from the National Archives of India

Foreign Political Department 1756-1938
Foreign Secret Department 1756-1938
Foreign Miscellaneous Series

B. Government Publications

General

Barnes, C.G., "Memorandum on the District of Bussahir and the Pacification of the Disaffected Portion of its Inhabitants", *Selection from the Public Correspondence of the Punjab Government* (Lahore, 1861), Vol. 5, No. 4.

Briggs, (Capt.) J.D., "Report on the Operation Connected with the Hindustan-Tibet Road from 1850-55", *Selection from the Records of the Government of India* (P.W.D.) (Calcutta, 1856), No. XVI.

Himachal Pradesh, Directorate of Eco. Statistics, *Facts and Figures of Kinnaur District* (Simla, 1962).

Himachal Pradesh, *Evaluation Study of Kinnaur* (Simla, 1971).

Himachal Pradesh, Directorate of Industries, *Survey Report on Handicrafts of Kinnaur District* (Simla, 1965).

Himachal Pradesh, Planning Department, *Sub-Plan for Tribal Belt 1974-77* (Simla).

India, Ministry of External Affairs, *Report of the Officials of the Government of India and the People's Republic of China on the Boundary Question* (Delhi, 1961).

India Planning Commission, *Report of the Study Team on Tribal Development Programme in Himachal Pradesh.*

Parmar, B.S., *Report on the Grazing Problems and Policy in Himachal Pradesh* (Simla, 1959).

Gazetteers

Himachal Pradesh Government, *Himachal Pradesh District Gazetteer. Kinnaur* (Simla, 1971).

Hunter, W.W., *Imperial Gazetteer of India, Vol. 5, Kunawar* (London, 1881), pp. 482-3.

Hunter, W.W., *Imperial Gazetteer of India, Vol. 5, Kunawar* (London 1886), pp. 361-2.

——, *Imperial Gazetteer of India, Vol. 14* (London, 1908), pp. 372-3.

Punjab Government, *Punjab State Gazetteer: Simla Hill States Bashahr State* (Lahore,1911).

Thorton, Edward, *Gazetteer of the Territories under the Government of the East India Company and of the Native States in the Continent of India* (London, n.d.), "Koonawur", pp. 517-18; "Bussahir", pp. 167-9.

Census

Census of India 1961: Himachal Pradesh—Fairs and Festivals: Kinnaur (Delhi, 1967), pp. 117-278.

Census of India 1961: Himachal Pradesh—Housing and Establishment (Delhi, 1965).

Census of India 1961: Himachal Pradesh-Kinnaur District-Census Handbook (Simla, 1966).

Census of India 1961: H.P.—Rural Craft Survey : The Art of Weaving (Delhi, 1968).

Census of India 1961: H.P.—Rural Craft Survey: Gold and Silver Ornaments (Delhi, 1967), "Kinnaur Design", pp. 95-100.

Census of India 1961 : H.P.—Scheduled Tables on Scheduled Castes and Scheduled Tribes (Delhi, 1967).

Census of India 1961 : H.P.—A Village Survey of Kanam (Delhi, 1965).

Census of India 1961 : H.P.—A Village Survey of Kothi (Delhi, 1963).

Census of India 1961 : Himachal Pradesh—A Village Survey of Nichar : (Delhi, 1966).

Census of India, 1971 : India Series 1—India Miscellaneous Studies : Ethnomusicology—Tribal Music : A Directory of Tribal Musicians of Kinnaur, (Delhi, 1972).

Census of India 1971 : Series I—India, Miscellaneous Studies Ethnomusicology—Tribal Music—The Music of Kinnaur (Delhi, 1975)

Ibbetson, Denzil, *Report on the Census of Punjab 1881* (Calcatta, 1883), 3 Vols.

SECONDARY SOURCES

General

Abbi, B.L., ed., *Urgent Research in Social Anthropology* (Simla, 1969). Problems of Western Himalayan Borderland, pp. 151-5.

Ali, S. Muzafer, *The Geography of the Puranas* (Delhi, 1970).

Baden-Powell, B.H., *Indian Village Community* (Delhi, 1972).

Bajpai, S.C., *The Northern Frontier of India* (Delhi, 1970).

Bajpat, P.V., Shan-Chien-Pi-Po-Sha, *A Chinese Version by Sanghbahadra of Samantrpasachika* (Poona, 1970).

Bose, N.K. *Some Indian Tribes* (Delhi, 1973).

——, *Tribal Life in India* (Delhi 1971).

Bose, S.C., *Geography of the Himalaya* (Delhi, 1972).

Budh Prakash, *Studies in Indian History and Civilization* (Agra, 1962).

Chakravarti, P.C., *The Evaluation of India's Northern Border* (Delhi, 1973).

Chatterjee, S.K., *Cultural Heritage of India* (Calcutta, 1958), Vol. I.

——, *Languages and Linguistic Problems* (Oxford University Press, 1943).

Chattopadhyaya, S., *Racial Affinities of Early North Indian Tribes* (Delhi, 1973).

Chopra, Pran, *On an Indian Border* (Delhi, 1964).

Coomaraswami, A.K. *Yaksas* (Delhi, 1971).

Cwell, E.B., ed., *The Jataks or Stories of the Buddha's Former Birth*, Vol. 4, (London, 1969).

Danielon, Alain, *Hindu Polytheism* (London, 1964).

Datar, B.N., *Himaloyan Pilgrimage* (Delhi, 1961).

Durant, Will, *Story of Civilization, Vol. I: Our Oriental Heritage* (N.Y.)

Fraser, J.B., *Journal of a Tour in the Snowy Ranges of Himala Mountains and to the Sources of Rivers Jamuna and Ganges* (London, 1820).

Fuchs, Stephen, *The Aboriginal Tribes of India* (Delhi, 1973).

Grierson, G.A., *Linguistic Survey of India*, Vol. I, (Delhi, 1967).

Govinda, (Lama) Aragarika, *The Way of the White Clouds* (Berkeley, 1970).

Guha, Amalendu, *Central Asia: Movement of People and Ideas from time Prehistoric to Modern* (Delhi, 1970).

Hamilton, W., *Geographical Statistical and Historical Description of Hindustan and Adjacent Countries* (London, 1820), 2 Vols.

Hedin, Even, *Trans-Himalaya*, Vol. 3, (London, 1909), pp. 340-414.

Hutton, G.H., *Caste in India* (O.U.P. 1961).

Ibbetson, Dr. Maclagan, *A Glossary of the Tribes and Castes of the Punjab and N.W.F.P.*, Comp. by H.A. Rose (Lahore, 1914-19), 3 Vols.

Jacquamont, V., *Letters from India. 1829-1932*, Tr. from the French by C.A. Phillips (London, 1936).

Joshi, S.R., *Our Borderlands* (Lucknow, 1964).

Lamoy, Richard, *The Speaking Tree* (O.U.P., 1971).

Mishra, D.P., *Studies in the Proto-History of India* (Delhi, 1971).

Mukherjee, R.K., *The Culture and Art of India* (London, 1959).

Negi, Thakur Sen, *Scheduled Tribes of Himachal Pradesh : A Profile* (Simla, 1876).

Neog, N., *Proceedings and Transaction of the All India Oriental Conference*, 22nd Session, January 1955.

Palis, Marco, *Peaks and Lamas* (London, 1939).

Pargitar, F.E., *Ancient Historical Tradition*, (Delhi,1962).

Rahul, Ram, *The Himalayan Borderland* (Delhi, 1970).

Randhawa, M.S., *Farmers of India* (Delhi, 1959). Vol. I, pp. 151-96.

Sachau, C., ed., *Alberunis' India* (Delhi 1964).

Sashi, S.S., *Himachal: Nature's Peaceful Paradise* (Delhi, 1971).

Singh, K. Suresh, *Tribal Situation in India* (Simla,1972).

Singh, Madanjit, *Himalayan Art* (London, 1970).

Snellgrove, D.L., *Buddhist Himalaya* (Oxford, 1957).

Stacton, David, *A Ride on Tiger: The Curious Travels of Victor Jacquamont* (London, 1954).

Thomas, P., Epics, *Myths and Legends of India* (Bombay, n.d.).

Thomson, Thomas, *Western Himalaya and Tibet* (London 1852).

Upadhayaya, B.S., *India in Kalidasa* (Delhi, 1968).

Vyas, S.N., *India in the Ramayana Age* (Delhi, 1967).

Zimmer, H., *Myths and Symbols in Indian Art and Civilization* (N.Y., 1953).

Kinnaur

Bailey, T.G., "Kanauri Vocabulary in Two Parts", *Royal Asiatic Society of Great Britain*, July 1910, pp. 659-705 and next issue, pp. 315-64.

——, *Linguistic Studies from the Himalayas* (London 1915) (includes Lower Kinauri and Chitkhuli).

Cumming, C.F.G., *In the Himalayas on the Indian Plains* (London, 1886).

Francke, A.H., *Antiquities of Indian Tibet*, 2 Vols. (Calcutta, 1913-1926)

Gerard, Alex., *Account of Koonawur in the Himalaya* (London, 1841).

Gerard, Alexander, "Travels in the Himalayan Country", *Selections from the Journal and Annual Register for British India and its Dependencies*, Vol. XIV to XXVIII, July 1822 to 1829 (Madras, 1875), Vol. II.

Hutton, Thomas, *Journal of a Trip Through Kunwaar, Hunguruny and Spiti Undertaken in the Year 1838.*

Joshi, T.R., *Grammar and Dictionary of Kanawari. The Language of Kanawar in the Bashahr State*, ed. by H.A. Rose (Calcutta, 1909).

——, *Notes on the Ethnography of Bashahr State, Simla Hills*, ed. by H.A. Rose in *J.A.S.B.*, September 1911.

Lloyd, Wm. and Gerard, Alex., *Narrative of a Journey from Caunpoor to the Boorando Pass in the Himalaya Mountain* (London, 1840).

McMahon, C.A., "Notes of a Tour Through Hangrang and Spiti", *Record of the Geological Survey of India*, XII, 1879.

Murray, Aynsley, A.C., *Account of a three Months Tour from Simla Through Bushahr Kunowar and Spiti to Lahaul* (Calcutta, 1882).

Panchamukhi, R.S., *Gandharva and Kinnaras in Indian Technology* (Dharwar, 1951).

Simpson, William, "Architecture in the Himalayas", *Transactions of the Royal Institute of British Architects 1882-83* (London), pp. 65-80 Illus.

Vander Sleen, W.G.N., *Four Months Camping in the Himalayas* (London, 1929).

Wilson, A., *Abode of Snow* (London, 1875).

शर्मा, (डा०) बंशी राम, किन्नर लोक साहित्य, (विलासपुर, 1976)

कौशल, रामकृष्ण, हिमानियों के देश में, (दिल्ली, 1967)

साँकृत्यायन, राहूल, किन्नर प्रदेश। (इलाहाबाद, 1957)

सिह, रघुनाथ कल्हणाकृत, राजतरंगणी, (वाराणसी, 1969)

हिन्दी विश्वकोश भाग 3 (वाराणसी, 1963)

साँकृत्यायन, राहूल, मध्य एशिया का इतिहास, (पटना 1956 भाग, 1)

Sanskrit Literature

Adi Purana of Acharya Jinasena (Kashi Jananpitha, 1963).

Mahabharata (Gita Press, Gorakhpur, Sambat 2009).

Manusmriti, Arthur Burnell Coke, Translation (Reprint, Delhi 1971).

Matsya Puraua, Notes by B.C. Majumdar, S.C. Vasu, H.H, Wilson, Bently, Wilford and others (Delhi, 1972).

Raghu, Vansha Kalidasa by Chhamlal Gyanchand Pathak (Varanasi, 1956).

Reg Veda, H.H. Wilson (London, 1888).

Shiva Purana (Delhi 1970).

Vishnu Purana (Gorakhpur, S. 2009).

Articles

Agnihotri, B.S., "Kinnar Desh", *Air Force Treking Society Records*, *1952-58* (Simla), pp. 45-54.

——"Legends of Kinnaur Desh", *Hindustan Times*, 20 September and 4 October 1959.

Ahir, D.C. "Temples of Kanum, the Pride of Himachal Pradesh", *Mahabodhi*, December 1971, pp. 44-45.

Cunningham, J.D., "Notes on Moorcrafts Travels in Ladakh and on the Gerard's Account of Kanawar including the General Description of the Latter's District" *J.A.S.B.*, Vol. 13, 1844.

Das, S.R., "A Note on the Kinnaras", *Man in India*, Vol. 34(1), January-March 1954.

Gerard, Alex., "Narrative of a Journey from Subathoo to Shipkee in Chinese Tartary", *Journal of Asiatic Society of Bengal*, Vol. II, 1842, pp. 363-91.

Goverdhan Singh, "The Origin of the Man and the Himalaya", *Himalaya Kalpadruma*, Vol. I, No. 1, 1964.

Khanna, H.R., "Trip to Kalpa Valley, *Times of India*, 28 February 1971, p. 7, col. 1.

Luthra, Indira, "The People of Kinnaur District", *Social Welfare*, Vol. 5(4), July 1958, p. 13.

Malik, Harji, "A Remote Village Tucked in the Himalayas: Sangla Feels the Impact of Change", *Sunday Standard*, 3 August 1975, p.7, col. 1-5 illus.

Mathai, Achamma, J., "Kinnaure: Belle of the Himalayas", *Social Welfare*, Vol. 10 (5), August 1963, p. 4.

Mehta, Swarnjit Kaur, "Kinnaur—Its Geographical Setting", *Everyday Science*, XIII (2), 1968, pp. 32-34.

Negi, Thakur Sen, "The Kanauras: Transformation of a Traditional Society", *Hindustan Times*, 28 June 1976, p. 9, col.1-5 Sketeh.

———,"Land of Ancient Kinners" *Cultural Forum*, Vol. XIII, Nos. 3-4, July, 1971, pp. 52-56, illus.

Parmar, Y.S., "Tribal Economy in Himachal Pradesh", *Indian Express*, 17 April 1972, p. 6, col. 7.

Patki, G.S., "In the Lands of Kinnars", *A.I.C.C. Economic Review*, Vol. X, Nos. 16-18, 9 January 1959, pp. 131-5.

Prabhakar, U.P., "Near Revolution in Kinnaur Life: Fruit Trees in every Household", *The Tribune*, 24 and 25 September 1975.

———, "Skating in Frozen Lake", *The Sunday Tribune*, 25 January 1976, p. 5, col 1-4, illus.

Raja Ram, "Through the Himalayas, from Simla to Tibet", *The Punjab Geographical Review*, 1(2), 1942.

Reeves, H.C., "Notes on the Mountains of Bussahir and Spiti", *Alpine Journal*, Vol. XXVIII, 1913.

Ramesh Chandra, "Types and Forms of Marriage in Kinnaur Village", *Man in India*, April-June, 1973, pp. 176-87.

Sadhu, J.N., "New Horizons Open for First Settlers of Himachal Pradesh (Kinnauras) *Indian Express*, 23 July 1972, Vol. IV, col. 1, illus.

Sawant, S.D., "The Kinnaur Valley", *Illustrated Weekly of India*. 10 February 1963.

Singh, Bulbul, "Kinnaur", *Illustrated Weekly of India* 23 March 1975.

Slewen, D.G.F.M., "The Way to the Baspa", *Himalayan Journal*, April 1929, pp. 67-79.

Temple, R.C., "Local Account of Bashahr State, Simla District", *Punjab Notes and Queries*, Vol. II, March 1885, p. 99 (Note 604).

Tribune Correspondent, "Wool Weavers of Himachal Pradesh", *The Tribune*, 2 July 1961.

Vaddya, K.L., "Kinnaur and Spiti: Before the Tragedy", *Patriot Magazine*, 16 February 1975, p. 2, col. 1-4, illus.

Index

Adi-Purana, 47
Administrative Changes (after 1948), 10ff
Adna malkiat (inferior proprietory rights), 103
Agriculture Department, 78
Agyatwas, 46
Ali, S. Fazal, 65-66
Allegiance, oath of, 21
Amar Kosha, 46,47
Animal Husbandry (pillars of tribal economy), 70
Anti-Corruption Unit, 96
Appellate Courts, 97
Arang-Se-dhori, 15
Arms Act, 96
Athwara (forced labour), 103
Ayurvedic Dispensaries, 123

Badhis, 43
Baildavang, 136
Baildayang, 33
Bakhal, 74, 75
Bandobist ceremony, 3
Bandobust, 136
Barnes, G. (Superintendent of Hill States), 64,101; settlement, 102, 142
Barter System, 88
Bashahr, 1, 52, 53, 57-61, 64, 78, 92, 97, 100, 119; estate, 105
Beas, 51
Bedhai, 28, 135
Bee keeping, 83,103
Begar (forced labour), 99-100, 105
Begnang, 27
Bennang-hachis (love marriage), 135
Bhaba Project, 144
Bharat Sewak Samaj, 123
Bharatavarsha, 50
Bhot rule, 137
Binanes, 37
Block Development Agency, 125
Bolzanu-Pona, 16
Bonyungabu Chashimis, 39
Border Roads Development Board (Deepak Project), 113
Borshimik, 33
Bose, 28
Brahma, shadow of, 46
Brahminism, 13
Bridge, Karcham, 5, 115; Rani, 116; Wire-rope, 116
Bridle paths, 115
British East India Company, 86, 138
Bruske, Rev. J.T., 119

Castes, 10ff
Celibacy, Concept of, 48
Chait Sankranti day, 28
Chamangs, 81
Chamba *Vanshavalies*, 54
Chamu-Kurti (woollen shirts), 15
Chamu-Sultan (woollen pyjama), 15
Chaukhat, 13
Chhangshimig, 28
Chhitkul, 15
Chhtrapati (title), 138
Chhubha (long coat), 15
Chigu (goat), 70
Chinese aggression, 111
Chini tahsil, 1; settlement of, 100; Tahsildar of, 107; Zamindars of, 103
Chironigns (during trance), 53
Chobas (itinerant village doctors), 122

Cholera epidemic, 101
Chuli Khali (fruit), 27
Climate, 6ff
Community Development Block 123; Agencies, 143
Community Development Programme, 125; life, 38ff
Credit facilities, 89
Crops, 75ff; Kharif, 74-77, 106, 139; Rabi, 74, 76, 106,137
Cutcha, 5

Dakhang Pashur (drink), 82
Dances, types of, 39
Darohi, 21
Dashongi, 94
Dev Purna, 53, 54
Dhara, 5
Dhaula Dhar, 1, 2, 4, 48
Dhori (woollen sari), 15
Dietetic items, 74
Dimtangohis (love marriage), 135
Do-Song Mang-Po-Je, 55
Domang Community (blacksmiths), 81
Dress and Ornaments, 15ff
Dubdhab or *Khuchis* (marriage by capture), 34
Dubwa (coin), 88
Dumtangshis (love marriage), 33, 34
Dunber Singh, 53
Durga Singh, 102
Dusrang, 13

Emerson, W.H. (I.C.S.), 63, 95, 103, 118
Excise Law, 96
Export, articles of, 87

Fateh Singh, 64
Festivals, 40-44, 154ff
Forest Lease, 64; Research Division, 78
Fraster, James Baithie, 19; British traveller, 69, 85-86

Gachhang, 133
Gamey, 17
Gampo, Tibet Sangtsen, 54
Gandha Madan Mountain, 49
Gandhi, Indira, 64
Gangajal (holy Ganges water), 135
Genealogy, 167ff
Gerard, Alexander, 69, 85
Ghanti, 16; Kinnaur liquor, 82
Ghoris (official), 94
Gomutra (cow's urine), 135
Gorasang (house warning), 13
Gotang (water-mill), 14
Gotra (clan), 11
Government Craft Institute (Kalpa), 86
Government Industrial Extension Centre (Textile) Nacher, 86
Gram Panchayat, 127
Guhyayas, Hatakdesh of, 49
Gurkha Occupation, 59ff; War, 61, 86, 93, 99
Gurkhas of Nepal, 59
Gyolang (celibate), 19

Habas, 122
Handicrafts, Cottage Industry and Natural Resources, 79ff
Har (type of marriage), 135
Hari, 34
Harun-Al-Rashid, 55
Hasgotang (hand-mill), 14
Hay, Lord William, 62, 101; assessment of 1856, 100; settlement in 1856, 142
Hemkuta Mountain, 49
High Court (Simla), 98
Himachal Pradesh, 1, 31, 34, 49, 65-66, 77-78, 85, 89, 96, 105, 108, 111, 113, 120
Himachal Pradesh Panchayat Raj Act, 1968, 98, 127
Hindukush, 50
Hindustan-Tibet Road, 8, 96, 111-15, 143
Home Guards, 96

Homskad (language), 10, 132
Horticulture, 77ff; source of income, 140
Household Utensils, 146
Housing Pattern, 12ff
Hukka thoti (*chilam*), 17
Hydro-Electric Development, 126
Hydro-electric Power Station (Nogli), 85, 144

Ibbeston, Denzil, 20
Indian Council of Agricultural Research (Kalpa), 77
Indian Revolt (1857), 101
Indian-Tibetan Border Police (GOI), 109; Security, 96
Improvement Measures (Himachal Government), 72ff
Information and Broadcasting, Ministry of (GOI), 109
Insecticides, 79
Intelligence Bureau (GOI), 109
International boundary, 130

Jadakaneits, 105
Jagirdars and Muafidars, 105ff
Jagirs, 105
Jagithang (chorus), 39
Jallianwala Bagh, tragedy of, 65
Janapadines, 49
Janekang (type of marriage), 135
Janetang or *Janekang* (legal marriage), 31, 33
Jangi, Thopan Project, 144
Jangiam dialect, 10
Janpada-ki-Pin, 51
Jatakas, stories of, 48
Jethang (primogeniture), 23
Jigich-Boba (younger father), 25
Jushis (love marriage), 33, 135

Kabul River Valley, 50
Kalidas, 47
Kalpa Line Road, 114
Kalyan Singh, 58
Kanchang, 23
Kanaur, Kanet Septs of, 203ff
Kanda, spirit of, 42
Kandas (kind of crop), 74-75
Kanets (clan), 11
Kanungo, 106
Kanyakubja Brahmins, 11
Karahads, 100, 102-3
Karali, 74-75
Karchham-Sangla-chhitkul Road, 8, 115
Kasturba Gandhi Smarak Nidhi, 123
Kayang, 39
Kehri Singh, 59
Khadi and Village Industries Commission, 124, 143
Khadra Dang, 114
Khads, 5
Khanda (wooden box), 14
Kharacha, 14
Kharif season, 78
Kharo suspension bridge, 114
Khaskis, 28, 135
Khillat (robe of honour), 138
Khuchis (type of marriage), 135
Kiar, 74-75
Kilba-Shaving Khad Road, 8
'Kimpurusha Khand', 49
King Banasur, 48
Kingship Terms, 150-51
Kinnaur, inhospitable terrain of, 9; *nautors* in, 103
Kinnaur District Co-operative Transport Society, 89
Kinnauri-Jangram, 10; women's ornaments, 16
Kinner Desh, administration of, 92; Kailash, 42; *kanthis*, 17, 131; tribe, 9
Kithen Vegetable, 77
Konals, 61
Kot Kangra, 59
Krachogmig (mundan ceremony), 29ff
Kublai Khan, 57
Kuninda tribe, 51
Kunzru, H.N., 65

Ladakh, 55-57, 81, 87, 88
Ladakh-Kailash-Trans-Himalayan Chain, 48
Lamiac Buddhism, 18
Lamas, 19
Lambardar, 94
Land Estates, Himachal Pradesh Abolition, 106
Land Reforms Act (1953), 106
Land revenue, 104; collection of, 106
Language, 10
Lawrence, Sir John (Chief Commissioner of Punjab), 62
Leo (Mahatma Gandhi Ashram, Hangrang), 120
Lord Shiva, 42-43

MacNabb, J.W., 102 ; settlement of (1874), 142
Magadha, Kingdom of, 50,
Mahabharata, 47, 49
Mahasu district, 1
Majomi (go-between), 31
Maligorkh (medium worshipper), 29
Mang-Song Mang-Tsen, 55
Manohar Das , 61
Mansukh Das Wazir, 62
Manu (great law giver), 47
Marriage, Hari type of, 38; settlement of, 30; ceremony, type of, 34; customs, 35
Matsya Puran, 46, 47
Maurya, Chandragupta, 50
Melup (superstitions), 22
Mieling (hearth), 13
Military and Border Security Forces, 118
Misal haquyat (standing record), 103
Mitchell, 103
Modern Civilization, advent of, 23; Kinnaur district, 100
Moghul Emperor Akbar, 138
Monasteries, 19
Mongoloid Origin, 105
Moravian Mission, 119, 122
Muafidars, 105
Muafis, 103, 105
Munshi Shamlal, 64
Musical Instruments, 17

Nachar link road, 114
Namgia Link Road, 8, 114, 116
Nango jo (Barley), 82
National Highway, 96, 112-15, 130, 143
Natpa (Pandra Bis Kalyan Ashram), 120
Nautar lands, 127
Nazrana (forfeit), 11
Nepal, 55; troops, 60
Newal (type of land), 74-75
Nomadic pastoral people, 92, 141
Non-cooperation movement (Simla Hill), 65
North East Frontiers Agency (NEFA), 108
Nunneries, 19
Nyaya Panchayats, 98, 142

Occupation, 7; agricultural labour, 7; cultivators, 7; fishing, 7; forestry, 7; hunting, 7; livestock, 7; quarrying, 7
Ochterlony, Sir Daniel (British Agent), 59
Oo-pagey (symbolic ritual), 32
Ornaments, 147ff
Orung, 11

Padma-bkai-t'anyio, 56
Pakpa (mat), 14
Panchratna, 37, 135
Panchayat Raj Act (1968), 127
Panikkar, K.M., 66
Parmar, Y.S., 65, 66
Parvataka (Himalayan King), 51
Parvatiya Adim Jati Sewak Sangh, 121, 123, 124
Pasham (fleece), 71
Patel, Sardar Vallabhbhai, 65
Patwari, 106
Pawari Konal, 61

Petech, 56
Phatta-Orang (dress), 15
Police Wireless Stations, 118
Poltu, 23
Polyandrous family, 23-25, 38; marriage, 32
Polyandry, Institution of, 25; system of, 25
Popular Revolts, 64ff
Population of the Kinnaur, 7
Postal Department (GOI), 109
Primary Health Centres, 123
Project, Bhabha, 126; Jangi-Thopan, 126; 144; Nathapa-Jhakri, 126, 144; Shongtong-Wangtu; 126, 144; Topan Powari, 126, 144
Project Implementing Committee (Kalpa), 124
Prithvi Narayan Shah, 138
Prosperity, boon of, 42
Public Works Department, 113, 133, 134
Puh tahsil, 131
Punjab, 70; Government, 65; Police Station, 96
Puranas, 48

Rang Foch Simant (muleteers) Cooperative Society, 89
Rahul Sanskrityayan, 70, 97, 137
Rai Saheb Mangat Rai, 95, 97
Raja Chatar Singh, 58
Raja Durga Singh (Solan), 65
Raja Kehri Singh, 58, 137, 138
Raja Mahendra Singh, 61, 62, 93, 99 100
Raja Padam Singh, 95, 97, 103, 118, 138, 142
Raja Sansar Chand (Kangra), 59
Raja Shamsher Singh, 63, 100, 138
Rajavansa (ruling class), 58
Rajputs, 11
Rajtarangini, 46
Rakchham, 115
Ram Bahadur Singh, 62
Ramayana, 47
Rampur tehsil, 1
Ran Bahadur Singh, 61, 65
Ranjit Singh, 60, 138
Rapa Valley Road, 8
Reet, custom of, 31
Regular settlement operation, 103
Rehja, 30
Religion, superstition, 188ff
Revenue Administration, 98ff; estates, 105
Rgyal-Bu-rin-chen, 57
Ribba-Rispa link road, 114
Rig-Veda, 46
Rituals, 26ff
River, Baspa, 5, 71, 117, 126, 130; Bhabha, 126; Chandra Bhaga, 48; Ganga, 73, 48, 50, 51; Jamuna, 48, 51, 60; Kashang, 130; Mulgaon, 130; Pejur, 130; Rohini, 50; Ropa, 130; Spiti, 1, 2, 54, 113, 117, 126, 130; Sutlej, 1, 41, 43, 48, 50-52, 54, 56, 59, 60, 71, 74, 75, 85, 98, 113-115, 126, 130, 137, 138; Tibetan, 137; Tirung, 130; Wanger, 130; Yamuna, 50; Yula, 130; Yulang, 130
Rogi (Thakkar Bapa Kinner Ashram), 120
Rohilkhand, 51
Rohru Tehsil, 103
Ropa (for Shyaso) Valley Road, 114
Roso, Captain, R., 86
Rural Industrial Training Institute, 86

Sajeran, 11
Sanad, 61, 62
Sangauli, treaty of, 61
Sansar Chand (Nalagrah Chief), 60
Saraddi Wazir, 61, 92, 141
Sat Khund, 57
Satkhand, 130
Scheduled Caste Community, 92
Schreve (Moravian Missionary), 80
Seemant Pradeshi Society, 123, 124, 143
Sessions Court (Simla), 98

Shajra Kishtwar (field map), 102, 103
Shamlal (Tahsildar of Nurpur), 62; settlement of, 101, 102
Shamsher Singh, 62
Sheep Breeding Farm (Karchham), 72
Shiva Purana, 47
Sholdang Khad, 114
Shukla Poksha, 21
Shumccho dialect, 10
Shuna Wazirs, 61
Shyaso Mehbar, 120
Sianachari System, 126
Sianas, 126
Sians, 126
Simla, 1, 6, 8; hill, 63, 103
Single-line administration, 107, 108
Skyid-Lde-Ni-Ma-Mgon, 56, 57
Sleen, Van Dar, 95
Small agricultural holding's fragmentation, 24
Social Service Organization, 123
Social Welfare Board (Himachal Pradesh), 124, 143
Sorang, 5
Special Security Bureau (GOI), 109
Staple food, 16, 133
States Reorganisation Commission, 65
Streams (for fishing), 71
Sucheran, 11
Sumdo mountain, 113
Surendra Shah (Raja of Garhwal), 63

Tahsildar, 106
Taiti Valley Road, 8
Takhshila, 50
Takkariya Jataka, 48
Tankari, 119
Tapra-Se-dhari (dress), 15
Tapru-Se-balzam-Pono (shoes), 18
Taranda-Baruna Khad, 116
Teg-Boba (elder father), 25
Temperate Vegetable Seeds, 77
Temple, tenants of, 106
Territorial Council (1957), 128
Thapa, Amar Singh (Gurka Leader) 59, 60
Thepang (woollen cap), 15
Tibet, 25, 57, 58, 73, 87, 88, 111
Tibetan Imperialism, 54ff, 137; Ladakhi-Moghul War (1681-83), 58, 138; territory, 4
Tikka Raghunath Singh, 61-63, 74, 97, 98, 100, 126, 138, 142; settelment of, 102, 142
Toshiming (to sit together), 17
Toshing, 134; *Konich*, 17
Tri-de-Songtsen (800-815), 55
Tri-de-Tsung-Sen), 55
Traut Farm (Sangla), 71, 73

Udai Singh, 59
Udanang (ceremony), 33, 136
Upper Hindustan-Tibet Road, 8
Utpala, 57

Valley, Baspa, 2-4, 6, 116, 130; Bhabha, 130; Chumbi, 112; Gandhara, 50; Gyanthing, 116; Hangrang, 3, 5, 107, 124, 131; Kalpa, 10; Kangra, 59; Ropa 2, 85, 130; Sangla, 4, 10; Spiti, 116, 130; Sutlej, 2-4, 6, 37, 111, 113, 114, 130; Taiti, 114; Tidong, 3, 116, 130; Tirasang-Subsidiary 116; Wangpo, 130
Vijay Singh, 59
Villages bordering Tibet, 118
Vir Bhadra Singh (Deputy Minister), 138
Vishnu Purana, 46, 47

Wiswas (degree of ritual purity), 11
Wangto, Bhabha Path, 115
Wazas, 11

Wazib-ul araz (village administration paper), 103
Wazir Chaturbhuj, 104
Wazir Families, 61
Wazir Fateh Ram, 60
Wellesley, Richard Colley, 138
Women, Social Status of, 23ff

Zadskar Mountains, 1,2
Zomos, 19, 29
Zung-Tee, 85